Leaders

How Top Innovators Can Help
Your Business Succeed
On a Global Basis

Leanne Bucaro

Manor House Publishing
www.manor-house.biz
905-648-2193

Library and Archives Canada Cataloguing in Publication

Bucaro, Leanne
 Leaders: how top innovators help businesses succeed on a global basis / Leanne Bucaro, author; Michael B. Davie, editor.

ISBN 0-9781070-2-0

 1. New products--Management. 2. Technological innovations--Management.
3. Creative ability in business. 4. Success in business. I. Davie, Michael B. II. Title.

TS155.8.B83 2006
658.5   C2006-903049-9

Cover photo of Earth courtesy of NASA Glenn Research Center

Copyright 2006: Leanne Bucaro/M. B. Davie/Manor House Pub.
Published December 15, 2006
Manor House Publishing Inc.
www.manor-house.biz
(905) 648-2193
First Edition. 144 pages. All rights reserved.
Cover Design: M. B. Davie

Manor House Publishes gratefully acknowledges the financial support of the Government of Canada through the Book Publishing Industry Development Program (BPIDP), Dept. of Canadian Heritage, for our publishing activities.

For:
My intellectual father, Gerald Ruryk, who always wins at Trivial Pursuit; and reminds me to work harder when things get tough; My mom, Judy Nayduk, for without her courage and encouragement I wouldn't be where I am today; My daughters, Amber and Hayleigh, for always being there and picking up the slack; Michael B. Davie for expert help in making my dream of becoming the author of this book a reality; Alan McLaren my business partner, confidant and other half of Infinity Communications, who was ready with encouraging words; ready to lend a hand. Without him, it just wouldn't be much fun.

About the author

Leanne Bucaro is passionate about helping businesses thrive and grow in a crowded marketplace. With over 20 years communications experience, Leanne has helped Fortune 10 to Fortune 1000 companies recognize unprecedented growth in their industries by taking a holistic approach to Public Relations, Communications and Sales.

Leanne and her business partner Alan McLaren, principals of Infinity Communications Inc., focus on assisting small to medium sized businesses that are serious about taking their business to the next level.

Leanne's first book, co-written with Alan McLaren, "PR Mentor" can be found at www.pr-mentor.com.

Leaders is Leanne second book, and a book she is passionate about – as the best way for new and growing business owners to learn is to emulate those that have already been there. "Success stories show us the path to a sometimes less travelled road."

About the Publisher

Founded in 1998, Manor House Publishing Inc. is today one of the most innovative business book publishers in North America.

For business leaders interested in creating a book, Manor House offers a virtual one-stop shopping experience, helping to develop books from concept to final edit version, and on to published and marketed hard copies. Manor House covers all production costs, revamps books for greater market appeal, places books in bookstores internationally, and makes books available to private buyers at discounts. Manor House titles benefit distribution networks providing worldwide market penetration, giving virtually all titles a global presence (see Manor House chapter in this book).

For more information, visit: www.manor-house.biz or email: mbdavie@manor-house.biz or phone: 905-648-2193.

Foreword

Leaders is an innovative and informative business book that's certain to educate and as it entertains with captivating bios of outstanding business leaders.

Author Leanne Bucaro adds another dimension to what might otherwise have been a standard collection of biographies on successful entrepreneurs. Her book is much, much more than that.

Leaders goes beyond business profiles to focus on those business leaders that can play key roles in helping other companies succeed – perhaps your own company.

Of particular interest are the chapters on business turn-around expert Robert Kubbernus and StreetSmart Marketer Michael Hepworth and many more.

In fact, every chapter in this book, every business profile, is an absolute gem filled with helpful and useful information and very good advice.

Other chapters of note including the lead, opening chapter on WSI relating this firm's pioneering work in simplifying the Internet and improving website designs to attract far more traffic.

Also of interest is the chapter on Mariposa Cruises – a truly unique and unforgettable venue for business meetings, seminars or virtually any corporate function you can think of.

There are also superb chapters on the author's own company, Infinity Communications, on Manor House Publishing and on CARSTAR, an industry leader that helps numerous companies keep their fleets running while settling insurance claims with lightning speed. A truly impressive mix of business leaders.

Perhaps the most helpful parts of this book are the superb Leadership Lessons concluding each chapter in which the featured business leader distils their advice in a list of unforgettable tips.

This information-packed book is a great read – and it's required reading for any business leader interested in achieving greater levels of success. Highly recommended.

- Michael B. Davie
Author, *Success Stories, Enterprise 2000* and *Winning Ways* series

INTRODUCTION

"The older I get the less I listen to what people say and the more I look at what they do."
-Andrew Carnegie

"You can have brilliant ideas, but if you can't get them across, your ideas won't get you anywhere."
-Lee Iacocca

If you own an independent business, or consider yourself an "entrepreneur," this book was written for you. My father, a wise and intellectual man, always advised me: "Be smart; learn from others' mistakes. Don't be stubborn and make them *all* yourself."

In working with many small and large businesses and working with leaders of Fortune 10 through Fortune 1000 companies – all the successful leaders agree – you learn through others' examples.

This inspired me and my business partner Alan McLaren to collect success stories of successful leaders in various fields.

You will discover personality traits from their early years – read some of their hardships and learn from their mistakes.

But most importantly, you'll be the recipient of their "leadership lessons." One of the most consistent traits in all of these stories is that the leaders understood that they couldn't do it all themselves. As the saying goes, "to be successful, a leader needs to work *on* their business, not *in* their business." I'm also happy to say that most of the companies featured in this book have been clients or are still clients of Infinity Communications.

As James Humes noted: "The art of communication is the language of leadership." Good leaders over communicate. Studies show that people need to hear a message at least seven times before it will stick. The first thing we do with our clients is Key Messaging. We have turned down clients that refuse to have key messages. Great organizations stick to their messages. Keep the message simple and don't be afraid to repeat it. Use multiple mediums – so your employees, customers, stakeholders hear it, see it, see and hear it, experience it… you get the point.

Look at who has already been where you want to go. How did they get there? Network in unrelated industries and listen to how people have overcome challenges in their industry – get creative. Read success stories – big companies and small companies that have had success. Apply their strategies. There are no bad answers – only bad questions.

In this book, you'll read the Robert Kubbernus success story – how he built several successful companies with ingenuity, creativity and simple strategies. He is not afraid to ask questions. He is also not risk adverse. Robert understands that strategies need to be dead simple. It is all about the execution.

Another of the chapters in this book is about Manor House Publishing – and how they help leaders in business get their message out with an innovative one-stop shopping approach in which books are written, edited, created, published and marketed. MHP helps share the message, and therefore not only helps the experts who write the books – but the people who read them.

WSI, Mariposa Cruises, marketing guru Michael Hepworth, CARSTAR and my own Infinity Communications, are other examples of how thinking outside the traditional box helps to differentiate you from competitors and achieve greater success.

A few quick tips before we get started: Don't be afraid to talk to people; network; ask questions; talk to your customers; pick your business partners well – because your partnership can be more intense than a marriage; hire great people; don't think you know it all. Be flexible. Be lean. As my partner Alan always says "take little bites off the elephant." If you try to eat it whole – you'll choke. Success is the sum total of all of your smaller decisions.

Growth isn't always easy – it can be downright painful – but well worth it. Most importantly, you can have a great business – but if the world doesn't know about it – it won't grow to the heights it deserves. Building brand and establishing yourself as an expert is key to garnering the attention you deserve. I talk about this more in the Infinity Communications chapter.

Now let's get started – hope you enjoy the book – and learn a lot from our collection of innovative and successful businesses.

- **Leanne Bucaro**

At a Glance: WSI:
We Simplify the Internet:
WSI Internet Consulting & Education.
Founded in 1995.

Leadership: Business: Within 10 years of founding, WSI has risen to be the leading Internet Solutions provider to small and medium sized businesses the world over, with more than 1,500 franchisees in 87 countries. WSI is ranked the No.1 Tech Services franchise for the fifth consecutive year. In 2006, WSI broke through to make the list of Top 50 Franchises in *Entrepreneur* magazine's annual Franchise 500 index. WSI today represents one of the most profitable proven franchise opportunities in the world. It's also the world's largest network of Internet consultants.

Leadership: Charity: WSI plays an impressive leadership role with corporate efforts, in co-operation with World Vision to eradicate child poverty and revitalize impoverished communities.

Key Strengths: Simplifying the Internet to help businesses and organizations, including charities, maximize benefits.

Services: World's top Internet consultancy network has Internet marketing, website and professional development, education.

Status: Private company, owned by a group of investors.

Employment: 125 staff at Mississauga world headquarters, plus more than 1,500 Internet consultants around the world.

For more Information:
Contacts: Ron McArthur, President; Marcia Scott, Vice-President, Product and Marketing; Sheetal Pinto, Manager, Public Relations and Communications.
Phone: 905-364-1348. Toll free (U.S.): 1-888-678-7588 ext. 1348
Fax: 905-678-7242.
Mail: 5580 Explorer Dr. Suite 600, Mississauga, Ont. L4W 4Y1
Email addresses for key personnel: rmcarthur@wsicorporate.com; mscott@torontohomeofficewsicorporate.com; spinto@wsicorporate.com
Websites: www.wsicorporate.com, www.wsiconsultants.com and www.wsimarketing.com

1

WSI

We Simplify the Internet

Mastering – not just accessing – the Internet is what separates the successful from the also-rans in today's fast-paced information technology-driven world.

With its ability to instantly share information/communication and propel mass marketing, the Internet has become a crucial ingredient in the success stories of countless large companies with the deep pockets necessary to pay for IT expertise.

Yet, many small to mid-sized firms are lost in cyberspace, hoping someone, anyone, will visit their lonely, uninviting, user-unfriendly website. But visitors are few.

While their competitors race past them on the Information Highway, they're still stuck in the Information *driveway*, fumbling about for those elusive keys to success.

They know they're missing out on enormous opportunities: There are more than 1.1 billion Internet users worldwide, most of whom make online purchases. Those purchases totalled $1.6 trillion in 2003 and today the total approaches $7 trillion. That's an enormous potential customer/revenue base. And most purchasers don't care about the size of the company they're buying from.

Those firms lost in cyberspace desperately need an Internet navigator and facilitator, a guide who will help them to maximize their online presence and effectiveness.

And they'll find such an expert in WSI, a low-key company that delights in giving its clients a high-profile on the world-wide web – and in as fast and cost-efficient a manner as possible.

WSI knows simplicity is core to unravelling the complicated nightmare that too many Internet novices find themselves mired in.

In fact, the WSI initials stand for: We Simplify the Internet.

And like the Internet itself, WSI has a worldwide presence with more than 1,000 franchise owners/consultants in 87 countries, providing Internet solutions to a gratefully profitable and growing clientele of small and medium sized businesses.

World's leading Internet franchise

WSI is also showing clear leadership in a highly demanding field. It's currently ranked the No. 1 Tech Services franchise ; amongst the Top 50 Franchises and the 6th Top Global Franchise offering international opportunities, by *Entrepreneur* magazine.

WSI is also the world's largest network of Internet consultants.

Although it's not a household name, there's no question that WSI today represents one of the most profitable proven franchise opportunities in the world, driven by demand for its services and the flexibility allowed franchisees to customize their businesses.

Perhaps most impressive is the simple fact that WSI is by far the leading provider of Internet solutions to the world. This is clearly the company to turn to for any business interested in tapping into the rich sales potential of the Internet.

Indeed, e-commerce sales, which rang in at $95.7-billion U.S. in 2003 are projected to reach $316-billion by 2010 – and to continue growing at an annual rate of 14 per cent, more than double the economic growth rates of Canada and the United States.

WSI command central can be found in an unassuming, modern Mississauga office building with a panoramic, bird's-eye view of Toronto's Pearson International Airport.

It's here, in the flight path of massive, passenger jets that you'll find the 125 people who serve as a support system for a vast, global network of Internet solution providers.

With the dynamic presence of powerful aircraft in flight serving as a backdrop, two key WSI personnel shared their company's own dynamic vision for helping businesses succeed.

WSI President Ron McArthur, and Vice President, Product and Marketing Marcia Scott concur their company is playing a vital if little-known role in helping small and mid-sized firms achieve growth and prosperity.

McArthur says WSI has tapped into a largely under-serviced, previously ignored market consisting of small to mid-sized firms needing Internet help.

"A lot of the larger firms have no problem spending a lot of money on Internet expertise – but the smaller ones can't afford it," notes McArthur, a middle-aged amiable man with a ready smile and handshake.

Reaching out to small to mid-sized enterprises

"The IBM type client firms were getting great solutions – but these were huge firms," McArthur notes. "There was really nothing out there for the little guy – and this represented an enormous market opportunity. The key was then to make these solutions economical and cost-efficient for smaller companies."

"We primarily sell to the small and mid-sized companies market – it's over 90 per cent of the business we do in any country," he adds. "And we've found that regardless of what country these client companies are in, they have very similar needs, including the need to focus on growing their business, doing what they do best while leaving other aspects, including the Internet, to the hired experts we can provide – we deal with many, many entrepreneurs."

"It's not that we're against promoting ourselves or that we haven't thought of doing that – we'd just rather promote our clients," McArthur explains. "That's our primary focus."

In assisting small to mid-sized enterprises, WSI employs its Internet Solutions Lifecycle System, a proven development plan designed to achieve maximum online success.

It functions as a circle of steps, including Internet business analysis to determine the best products and services, functional design, building, testing, launching and managing the results, which in turn is followed by an updated Internet business analysis to repeat this quality management cycle to ensure production of the best products and services.

WSI also has a system that provides the kind of access to products, services and solutions normally reserved for much larger firms.

Known as MAPS (Master Production System), this system constantly brings in the best products from WSI's production partners, employing WSI's Internet Solutions Lifecycle System™ and resulting in production and development of superior products at reasonable prices.

MAPS also draws sales leads from lead-generation partners, feeds in Internet solutions from authorized suppliers and draws hosted web solutions from hosting partners.

All of this is utilized in the development of Internet solutions, which are then accessed by WSI's Internet consultants and passed on to their clients, the small to mid-sized enterprises, which then adopt and utilize these advancements to achieve greater profitability.

Custom-made Internet solutions

McArthur says WSI leads the way in providing expertise tailor-made to individual companies via local consultants, with a goal to better exploiting the world market.

"This is a unique Canadian company, offering a global presence with a local touch so that the customer can better compete in a 24-7 marketplace," he explains.

Scott smiles in response. "We essentially run our own economy inside 87 different countries," she notes.

"One of our strengths," Scott adds, "is that there are over 1,500 WSI Internet consultants in the field in these various countries."

"Our consultants are very familiar with the idiosyncrasies of each market and they all have the latest information on leading edge technologies and how they can best be applied for each client's benefit."

For example, Scott notes stringent efforts are made to select an affordable and effective e-commerce product that's the best in its class.

To find that product solution, WSI subjects competing products to a rigorous multi-step selection process "and we then award the opportunity to the best-in-class company," Scott explains.

McArthur observes that WSI goes beyond simply making the best products and solutions available at very affordable rates.

"Our consultants work closely with the customer to really understand their needs," he points out. "The Internet consultants aren't there to sell a given product – they represent all solutions custom-built by our production centres. They're there to find the right products for the customer after first going through a thorough needs analysis of the client."

"Then," McArthur adds, "the solution designed for the customer is custom-assembled from various components."

"There's a customized look to the design of the website," he notes, "and every project is priced to the needs of the customer. We've partnered with our Internet consultant franchisees – our Internet business experts – who deliver Internet products and services that best fit a given business's needs and budget."

"Our consultants look for ways to reduce costs, build revenue and grow the customer base by introducing cost-effective e-business tools," he adds.

Modular components key

"The key," McArthur notes, "is to make the components of the system modular so they can be mixed and matched, and we've done that."

"It's an effective and affordable solution," he adds.

How affordable? McArthur says the Internet consultants can charge as little as $1,200 to $1,500 U.S. to get an effective 2-page website established for a start-up business – and that includes consulting time, maintenance and hosting services.

The average cost could be as low as $5,500 U.S. to have a multi-functional e-commerce site installed.

And some firms find they can finance this without feeling a pinch, simply by reducing the size of their Yellow Pages phone book advertisement.

But there are more compelling reasons than cost for using WSI's services:

Beyond the savings, there's ease of use of the technology: "It's set up so the customer can keep information updated as easily as working with a Word document," McArthur points out.

"There's no having to fool around with html or coding of any kind," he adds. "The customer can easily change content as frequently as required."

McArthur notes WSI will also perform WSI Web Scans™ analyzing the structured content and effectiveness of the website and the number of hits, or visits, it attracts.

The Internet consultant can then apply his or her expertise to achieve search engine traffic maximization. WSI consultants also know that getting a client's press releases to be among the first items that appear in a Google search is a critical step in world-wide Web recognition of a business.

Instant impact

"It's an instantaneous hit," Scott smiles. "We're speed of light. And it's vitally important to make it easy for people to find your website. E-commerce, selling your products and services online is like having a store that's open 24 hours a day, seven days a week, all year, every year."

Scott notes there are numerous advantages to this approach: "We make it easy to instantly advertise specials, take purchase orders automatically, track inventory and respond to your customers' needs without having to hire a lot of staff."

WSI provides proven and effective expertise in providing a range of valuable Internet-related products and services, including: search engine maximization; website creation, development and enhancement; increased numbers of hits – or visits – to websites; set-up of e-commerce capability; marketing solutions, e-learning and staff-training systems, video email; web conferencing; multimedia, including flash and other attention-getting tools; and, e-marketing techniques and practices to increase the customer-base.

And there are long-term benefits to going with WSI.

"The biggest problem a company encounters elsewhere is that they'll hire a web designer who builds them a site and then walks away, never to be heard from again," McArthur notes. "They're also often trying to sell a group of products that may not best serve the client's needs," he observes. "Off-the-shelf website packages and do-it-yourself site builders are often just shortcuts that end up costing you money and leaving you with little support after installation."

"In contrast," McArthur adds, "we offer customized solutions and we're there for the long term."

"And our approach from the start," he elaborates, "is to look at future upgrade capabilities and potential so that the system can grow as the customer's business grows. In short, helping business owners make money while they sleep; for the life of their business."

McArthur notes WSI's proven system increases the profitability of customers and franchisees by "providing web-related products and services that deliver ongoing results."

Local expertise provided

"We can also provide an Internet consultant who is in the same city or region as the customer," McArthur points out. "Our research has shown us that companies want contact with an IC who is local to them, again in contrast to many of the consulting services out there."

Another distinction is that when a small business executive calls WSI for help, it's unlikely he'll be talking to an inexperienced Web designer or amateur babbling away in incomprehensible techno-speak.

McArthur and Scott, the rest of the 125-member headquarters support staff and the entire network of more than 2,000 Internet consultants are all of various age groups and backgrounds, with the younger members sensitive enough to the needs of older business people to avoid jargon in favour of plain talk.

In fact, many in WSI are career-change business people who know exactly where the business client is coming from, what he or she is going through, and how to best find solutions to their Internet problems – as they've encountered similar problems themselves.

They serve as experienced and astute business advisors focused on providing Internet solutions.

"Our people come from all walks of life and from all manner of backgrounds and work experiences," notes McArthur, a former coffee company executive who has gone from java to, well, java.

McArthur had started out as a chartered accounted for the Red Carpet Coffee Services in 1977, then the largest supplier of coffee services to offices, and later acquired by the Hudson Bay Company. McArthur served as company president from 1983.

The company was acquired in 1994 by Van Houtte, a Quebec-based coffee roaster founded in the early 1900s by Dutch immigrants.

McArthur stayed on as president until making the career change to WSI in early 2004 where he has continued to focus strongly on superior customer service and the development and deployment of practices for the company's franchise operations.

With a strong reputation as a leader and team-builder, McArthur was named WSI President in the fall of 2004.

Customer-focused growth

Scott, Vice President of Product and Marketing, joined WSI in May 2004 after a lengthy corporate career marketing financial products for TD Bank, American Express, Manulife and North American Life.

As the former Assistant Vice-President of Strategic Marketing for Manulife Financial's Affinity markets business unit, Scott developed a wealth of business-to-business and business-to-consumer marketing expertise.

A leader in the field of direct marketing, she's helped WSI significantly expand its product and service sales to small to medium sized business clients around the world.

She credits the entire management team for providing the sound mix of skills needed to help WSI enhance its strong commitment to achieving market expansion via customer-focused growth.

"We bring practical skill sets – understanding customers, fulfilling their needs, simplifying business arrangements, making the Internet easier to understand – to the WSI family," Scott explains.

"And we really are a family – consisting of people from all over who have come together with the common goal of helping others succeed by making the Internet easier to understand and use," she adds.

Both she and McArthur were drawn to WSI's innovative, entrepreneurial approach and determination to help smaller businesses succeed.

"WSI was built on the entrepreneurial spirit – it's something we understand, and it continues to drive our growth," says McArthur.

WSI's dominant Internet consulting position on the world stage is all the more impressive when it's recalled that the company isn't all that old. Its origins only date back to the mid-1990s.

This world-leading company was founded in 1995 by Dan Monaghan, who envisioned the full scope, power, benefits and advantages inherent in the Internet as a vehicle for achieving business success.

Dan also saw a substantial untapped business opportunity: bringing full Internet capabilities to small to mid-sized firms that were then being overlooked entirely.

Granted, the large IBM type client companies were getting great Internet solutions provided on a consistent basis – but there was nothing in the way of such services for smaller firms.

Given the sheer numbers of small businesses – the leading source of employment and economic growth in most countries – this represented a vast, untapped market.

Making Internet expertise affordable

Yet, given the more limited financial resources of smaller firms compared with their larger counterparts, the key was how to make access to Internet expertise affordable, cost-efficient and a sound return on investment.

The modular concept was developed to provide maximum flexibility and lower investment while the production facilities and global franchise network provided the business mass needed to attract bulk discounts, further aiding affordability.

And the network of franchisees has gone on to deliver literally thousands of profitable Web solutions to businesses around the world, leveraging WSI's international production capacity to enable success at the local level.

McArthur says the importance of the Internet to business success cannot be understated.

"If you're going fishing, you want to fish where the fish are," he explains. "It's the same in business – you want to be where your potential customers are."

"Now consider," McArthur adds, "that growing numbers of consumers are turning to the Internet as their primary source of research information for whatever products and services they're interested in buying. If they can't find you, it's a safe bet they will find your competition."

"So" he concludes, "you need to ask: Can I afford to lose that business? Or, should I be taking steps now to improve my Internet presence, stay competitive in a changing market, reach more customers and increase sales?"

"Using a WSI Internet consultant to achieve this will help you reduce your investment in time and money more effectively than if you attempted a do-it-yourself website or tried customizing a ready-made package."

Customers play vital role

McArthur also notes that WSI customers play an active role in the process: "You provide the content and review customized design concepts so that your website design is something you'll be proud to show your own customers."

"The WSI consultant will remain available to provide ongoing expert advice and guidance," McArthur points out.

"And our hosting systems are housed in secure facilities and monitored 24 hours a day," he adds.

"We have strict security policies to ensure your data is kept safe and our software and hardware partnerships with Microsoft and IBM achieve the highest levels of reliability. This is superb customer support from every standpoint."

The ABCs of Internet success

McArthur notes WSI's full range of products and services constitute a whole that is far more than the sum of its parts, enabling clients to truly become masters of the Internet.

"We often refer to what we offer as the ABCs of Internet success," he adds.

McArthur explains:

A stands for affordable, advanced technology – e-commerce, online collaboration, targeted email marking, fast automatic delivery of information to customers and suppliers.

B is for building targeted traffic – an absolutely crucial element as search engine optimization, increased site visitor traffic, linking programs and online marketing campaigns are vital to success, with four out of five small businesses citing an effective Internet presence as a key component of their firm's prosperity.

C is for converting more customers online – including website maintenance, content updating and reporting, Internet marketing campaign strategy and management, analysis and advice for online business strategies, e-learning programs and a business focus to your Internet presence.

Fighting child poverty

WSI is also showing remarkable leadership as a major contributor to many worthy causes.

One example: WSI consultants often provide video email services to overseas soldiers, making their time serving their country more enjoyable by putting them in touch with loved ones, particularly during holiday seasons and on occasions like Valentine's Day.

But perhaps the best example is the outreach support program WSI provides to assist the World Vision charitable organization. WSI uses its economies of scale and huge Internet presence to assist World Vision's efforts to help ordinary – yet remarkable – people sponsor eligible impoverished children in Third World villages around the world.

"WSI has the franchise industry's highest ranking on Google – we get over half-a-billion hits a month throughout our global network," McArthur notes, "and this strong Internet presence enables us to help worthy causes on a worldwide basis."

"Making poverty history is within our collective capability," McArthur elaborates. "At WSI we feel particularly close to the plight of disadvantaged children around the world. That's why our Youth Outreach Program is the centerpiece of our corporate sponsorship and fundraising."

Among the communities helped by WSI is the water-starved community of Ventanilla in Peru.

WSI established a sponsorship program through World Vision in 2003 that involves sponsoring all eligible children in the village, giving them nourishment and education, raising their families' standards of living and helping the whole community work towards self-sustaining prosperity.

WSI is also supporting the building of infrastructure in the village to provide access to clean water and adequate sanitation.

Further support is being given to the children in Thailand and the Philippines as WSI employs its global presence to assist young people around the world.

WSI is also using its Internet expertise and industry connections to get other corporations and individuals to join the fight to end child poverty. For every child sponsored by another firm or individual, WSI is providing matching sponsorship of additional children. Those wishing more detailed information, should visit: www.makechildpovertyhistory.org

McArthur notes WSI's multifaceted approach to helping businesses and impoverished communities succeed is very much in keeping with its corporate philosophy.

"We believe strongly in giving back to communities, but it goes beyond our charitable work in striving to eradicate child poverty," he explains.

"Our business-help focus is core to what we're all about: Small to mid-sized businesses create the majority of jobs just about anywhere in the world – they're by far the leading source of employment and economic growth," McArthur notes.

"If we can help them achieve greater success, that means more jobs, more wealth generated, more prosperous communities and stronger economies to the ultimate benefit of everyone."

Services offered to small to mid-sized firms include:

- **E-commerce:** utilizing WSI's advanced database technologies to allow the easy adding-on of items, descriptions and pricing to form online catalogues. This also includes the virtual Shopping Cart feature that's a familiar sight on professional sites such as amazon.com – it keeps a running total, calculates taxes and adds in shipping costs. Credit card processing at online check-out is another available e-commerce feature. You can also utilize banner advertising, discounts and other promotions. The system also tracks orders and manages inventory. The convenient features make it easy for customers to shop securely on your website around the clock. It's open for business 24 hours a day, seven days a week.
- **Website Hosting:** This is business-grade website hosting at its best, with state-of-the-art security systems to protect your data, systems monitoring on a 24/7 basis, no limits on disc space or bandwidth usage, Smartmail email technology, traffic filtering, virus and spam protection, unbeatable service availability – all at very reasonable rates.
- **Web Conferencing:** This includes a selection of best of breed online conference tools that transform PowerPoint presentations into multimedia shows that can be viewed anywhere with an Internet connection – no special software needed. These Web conference tools are terrific for audio-visual demonstrations, for providing customer help and technical support, for showcasing company forecasts and briefings, or for adding audio-visual impact to sales presentations. It also ultimately pays for itself in savings from reduced

travel expenses, greater work flexibility, enhanced ability to share files and data, and the ability to bring together key players into one Web conferencing centre – a private meeting room branded with your company logo and with real-time visual and voice communication.

- **Email Marketing:** An inexpensive – about the lowest cost of any advertising medium – yet highly effective means to promote products and services. It allows for target marketing in an efficient manner and yields fast and high response rates. WSI maximizes your email marketing effectiveness via search engines, the main method consumers use to find what they're looking for online. WSI has also introduced video email technology allowing firms to send customers an infomercial to their inbox and provide effective, impressive visual communications during online communications. As well, being listed and attaining a top rank for your industry's keywords can make a huge difference in consumer traffic. This is a proven way to connect with prospective customers at the exact moment they're researching what product or service to buy. Having your firm's name pop up instantly during such a search drives up consumer traffic, raises brand awareness and gives your company a strong advantage over competitors. Not having such an effective Internet presence is tantamount to handing customers over to competitors.
- **E-Learning Solutions:** This includes a self-guided learning approach providing access to the system anytime from home or work, allowing the learner to progress at his or her own pace, resulting in a more enjoyable and effective learning experience. Course outlines are displayed in an intuitive graphical format to clearly show the different course modules and paths to completion. An online messenger feature enables direct communication with the teacher and fellow students, creating a community of people helping one another. A notebook assistant feature

allows learners to take notes and save information and a progress report feature provides for ongoing assessment of advances made. A digital certificate can be personalized and awarded to learners on completion of a course. This is a highly effective teaching approach for everything from business manuals to full content-rich curriculum. This system efficiently creates courses and reports while offering huge savings: No need for training staff salaries or classroom facilities or time lost to transportation as it reaches employees in the most remote locations. Private branding can be included in the system as can Flash and other multimedia elements.

- Internet education services also include critical knowledge to protect firms from digital attacks by hackers by employing the right software and security systems. WSI educates clients to be prepared for hacker invasions that in the past have cost businesses tens of billions of dollars in damage. Indeed digital attacks are now believed to have caused in excess of $250-billion worth of damage, worldwide.

Leading Franchise Opportunity

WSI is ranked the No. 1 Tech Services franchise for the fifth consecutive year. In 2006, in 2006, WSI has broken through to make it into the list of the Top 50 Franchises and the Top 10 Global Franchises in Entrepreneur magazine's annual Franchise 500 index. WSI today represents one of the most profitable proven franchise opportunities in the world. So what's involved?

Top 10 reasons for applying for a WSI franchise:

1. **Cost and territory:** For $49,700 U.S. a WSI Internet Consultant franchise can be purchased that provides a license to operate in a specified territory – usually a section of a city – with a population base of at least 100,000 people. Rights are non-exclusive given the borderless scope of the Internet. A worldwide franchise license to provide Internet solutions on an international

basis can be acquired for $137,000 U.S. Small investments for potentially large and ongoing returns.

2. **Residency:** The franchisee must reside in the territory for which they've been granted a franchise license to operate. This is to the franchisee's advantage as WSI studies have shown clients desire access to a *local* Internet consultant.
3. **Flexibility:** The franchisees decide the scope of their business and how they want to operate in their given territories regarding the range of services and time commitment they wish to provide.
4. **Support:** WSI provides full support services, including access to the products it develops, expert advice and business expertise, serving as a facilitator to effectively enable each franchisee's success. WSI support staff works closely with the franchisees, developing successful long-term relationships of mutual benefit to WSI, the franchisees and the customers. McArthur notes: "We strongly sense the enhancements we provide are really appreciated by franchisees and ultimately, customers."
5. **Training:** Franchisees are trained, certified and supported as a WSI Internet Consultant through 4-6-week pre-training programs using e-learning software, followed by an intense one-week training course at WSI's head office. A business coach is then provided to help put best practices in place during the first 90 days of operation. The extensive training program ensures the IC can serve customers as a business advisor specializing in making companies more profitable via cost-saving and revenue-generating Internet technologies as applied to websites, e-commerce (selling online) e-learning (on-line training) and online marketing.
6. **Minimal Overhead:** There's no need to lease a storefront – you can operate your franchise comfortably from your own home office. WSI provides you with a WSI branded franchisee website presence at no added charge to assist in promoting your business online.
7. **No Inventory:** You're offering Internet products and services so there's no need to purchase inventory that would take up space and depreciate in value as it sat on

shelves. And the only equipment you need is a laptop computer with basic business software. Your added start-up costs, beyond the initial franchise fee are minimal. As with any franchise, there are minimum monthly franchise fees plus fees based on sales, making them predictable and manageable.

8. **Cost-effective Operation:** WSI uses the Internet to maximize your profitability, reduce costs and help you communicate with clients and venders more efficiently. WSI's MAPS enterprise business system lets you operate securely from any location with an Internet connection for added flexibility.

9. **High-growth Market:** There are more than 1.1 billion Internet users worldwide, most of whom make online purchases, and the total value of goods and services transacted over the Internet is expected to exceed $7-trillion U.S. in 2007. That's a vast potential customer/revenue base. WSI's support structure is designed to sell goods and services on the Internet, making a WSI franchise an excellent opportunity to take part in a growth industry.

10. **Proven Business Model:** In addition to training in the Internet Solutions Lifecycle developed by WSI and ongoing training opportunities, complete access to WSI's private e-Marketplace, and a secure log-in to the enterprise business system, you'll also receive access to the WSI Intranet containing pre-designed marketing materials, forms templates and other tools to help you promote your business online. It's all part of WSI's proven business model that has helped many WSI consultants and their clients achieve great success.

Leadership Lessons:

1. Find an untapped market and service it. This is the approach WSI has taken to providing Internet expertise to small to mid-sized enterprises. Prior to WSI, this degree of expertise and service was normally only available to large IBM type client firms. This lesson, and the others listed here, can be applied to numerous other marketplaces. Reaching out to overlooked markets can yield rich dividends for you and your clients.

2. Create and develop ways to provide the highest quality products and services to your customers at the lowest possible costs to overcome client affordability issues and capture market segments previously lost due to client cost concerns. This approach inevitably generates increased business volumes, which in turn allow for increased production, bulk buying and economies of scale that enable reasonable pricing and significant profits.

3. Develop a highly effective website to fully promote and sell your products and services to a vast marketplace. Be sure the search engines can bring visitors to your site and ensure your site is attractively presented and user friendly.

4. Make your Internet presence as strong as possible, fully utilizing the Internet and your website as a tireless sales force operating 24 hours a day, seven days a week to raise your profile and generate sales revenue on your behalf.

5. Give your website full e-commerce and e-marketing capability to drum up as much business as possible. Realize that the costs of hiring WSI consultants and putting needed systems in place is minor compared to the benefits. Realize too that the full costs involved are often less than an advertisement in the phone book and that every day that you don't have such capabilities in place is costing you money in terms of lost sales and business you've failed to capture.

6. Achieve success by helping others achieve success. Position your company as the go-to firm for expert advice and superb products and services that help other firms succeed.

7. Don't just sell products and services – assess each client's needs and match them up with those products and services that best suit their needs. Adopt a long-term approach that ensures whatever you sell a client today can be added on to tomorrow as the customer's business grows and requires updated and expanded products and services.

8. As you achieve success in a financial sense, strive to achieve equally satisfying success in helping the less fortunate help themselves. WSI's work with World Vision fighting child poverty and revitalizing Third World villages provide good examples of this approach.

9. Be there when your customers need you. Don't simply make a sale and walk away, be available to provide follow-up assistance and advice. This approach is also crucial for gaining trust, developing repeat business, follow-up sales and referrals.

10. Accept and embrace opportunity. When opportunity knocks, open the door. Develop goals based on something you're good at and that you enjoy doing. Play to your strengths. Think about what you want to achieve and set about working hard and smart to make your dreams a reality.

At a Glance:

Robert Kubbernus
Balaton Group Inc:
Corporate Architects

Leadership: With 20 years of expertise in financial work-outs and turnarounds, Robert Kubbernus (DOB: Oct. 23, 1959) leads the Balaton Group in execution, opportunity identification and strategy. With a proven track record in Europe, United States and Canada, Kubbernus has developed the key business, political and capital market relationships needed to build global organizations. Kubbernus has delivered more than $850 million in both debt and equity to special situation companies through his investor network, and has provided strategic advice as a board member to numerous micro-cap and small cap public and private companies. Noted achievements include nomination: Ernst and Young Entrepreneur of the Year; named on E&Y list of Top 50 Most Influential People.

Key Strengths: Turning around troubled companies to assist them in achieving their full potential and maximize shareholder benefits.

Services: Balaton remedies corporate market and capital issues, repositioning the business fundamentals, and then bringing the company back into an industry-leadership position.

Status: Private company, owned by Kubbernus and investors.

For more Information: Contact: Robert Kubbernus, President
Global headquarters: 152 King Street East, Suite 400
Toronto, Ontario, M5A 1J3
Phone: 416-366-5702 or 416-366-4556
Fax: 416-366-8273
Web: www.balatongroupinc.com
Email: robert@balatongroupinc.com & info@balatongroupinc.com

2

Robert Kubbernus and Balaton Group
Turning Bad Situations Around

"There are no rules – just rough guidelines"
- **Robert Kubbernus, President, Balaton Group**

Robert Kubbernus could scarcely believe the crisis unravelling in British Columbia. Radium Hot Springs Lodge at Kootenay National Park appeared to be in danger of sliding down the side of the Rocky Mountains. This once glorious mountain retreat located directly at the Park gates of Radium was now on the verge of ruin.

It was the Christmas season in 1994. The main water supply pipe running above the mountainside hotel had broken, flooding the entire building and down the slope directly above the Trans Canada highway. Parks Canada feared that the 30-year-old treasured hotel would be washed off the mountainside onto the Trans Canada and into the Sulphur Hot Springs. The environmental impact would be substantial, not to mention that the park gates saw over 9 million travellers each year with many stopping to enjoy a soak in the hot sulphur pools, owned by Parks Canada, that were located directly beneath the lodge.

Kubbernus with his expertise as a corporate turnaround expert was brought in on behalf of both the hotel operator and the lender RoyNat to help the troubled company get back on its feet when the flood hit. Kubbernus quickly realized the situation demanded more than financial know-how.

Disaster Control

"It was a matter of immediate disaster control at one of the worst times of the year, Christmas – I needed to get the right people in place immediately."

He called in a business colleague who owned a large engineering and construction firm in Alberta. The colleague stabilized the hotel structure and took control of the site while the restructure plan was finalized. This 66-room hotel and lodging site was crawling with insurance companies, Parks Canada officers, bankers, and of course, lawyers. Yet the core problem was a business fiasco.

"The site was being leased from Parks Canada and it was a landmark hotel steeped in history, great for romantic getaways, and only a two-hour drive from Calgary and its natural attributes were being squandered," Kubbernus recalls.

"Like most hotels, it supported several departments – maintenance, cleaning services, cuisine, tours, heli-skiing and so on – to keep it running smoothly. The current operator did not have enough working capital to support an operation of this type, let alone the new challenge of up-front capital to pay for repairs. Knowing that an insurance claim in the millions of dollars would take years to process, the current operator stood no chance to recover. Prior to the flood the operator was failing and kept cutting back, hoping the cash flow would pay for up-keep and maintenance that really couldn't wait. The building was in complete disrepair. It's a business, not just a piece of real estate, although the owners weren't treating it that way."

The hotel owners were also several months (more than $400,000) in arrears on their mortgage payments, which made it painfully clear that a negotiated receivership of sorts would be required. Kubbernus dismissed the operators, shut down all operations and set to work on the new business plan, financial turnaround and reconstruction program.

The principals of the engineering and construction firm retained to stabilize the building were also real estate developers with a portfolio of apartment buildings, motels and commercial properties. So, Kubbernus began negotiations between RoyNat, the displaced operator, the engineering and construction firm, the insurance company, Parks Canada and the potential acquirer. "Keeping conflicts of interest in check was one of the tough challenges" Kubbernus recalls.

From Potential Loss to $8 million Value

With the restructure, reconstruction, insurance settlements and receivership now well in the past, this old Rocky Mountains getaway has come from what was a $4 million potential loss to a current value of over $8 million today. A financial crisis was resolved; an ecological disaster averted.

The old operator was removed without financial ruin, RoyNat mitigated its mortgage and received 100 cents on the dollar, the hotel has been restored to its glory days and the new owner has received a very nice return for taking the risk at the beginning.

This rocky experience in the Rocky Mountains was one of the more dramatic moments in a career spent turning around seriously troubled companies, which involved a multitude of stake holders and competing interests. "Complexity and trouble is my employer," says Kubbernus.

Indeed, the expertise Kubbernus acquired over the years has made him the go-to guy for fixing the seemingly unfixable; for making right those firms that have gone tragically wrong. He's now CEO of Toronto-based Balaton Group Inc., which specializes in turning troubled companies around via careful restructuring and revitalization of the corporation. Kubbernus is often able to not only get the troubled company back on track, but frequently achieves impressive returns for the investors and restores stock values for public companies.

How Kubbernus became a leading expert on corporate turnarounds is a story that begins in his youth.

Father's Entrepreneurial Spirit Inspirational

Born in Edmonton but raised in Calgary, Kubbernus recalls an upper middle class, "self-directed" childhood in which his parents, Vern and Vera Kubbernus, expressed their confidence in him and "never held me back or tried to reign me in to spare me from a failure. I was always free to try new things and see how I did with them. Nor was there ever any pressure to work in the family business. I made my own life decisions and was free to try and succeed at anything I put my mind to."

He was also influenced by the entrepreneurial spirit of his father, a partner with JK Campbell and Associates, a construction company specializing in heating, ventilation and air conditioning systems for major corporations – such as oil companies – and large institutional customers.

At age 16 Robert Kubbernus left Alberta and moved to British Columbia where he completed high school. He was eager to make his own way in the working world and to meet up with his girlfriend who became his first wife two years later (they have since divorced; their son, Christopher, is now a Balaton Group executive).

Kubbernus returned to Alberta at age 18 and worked for the JK Campbell company until he was 22, completing three years of a four-year journeyman sheet metal apprenticeship program during this time through Southern Alberta Institute of Technology but gravitated to the business, management and design side of the program.

"I worked during the day for the 800 person firm JK Campbell and started my own firm at night doing residential jobs," he recalls, noting this did not present a conflict of interest as JK Campbell was exclusively involved in large commercial projects. The company largely ignored smaller jobs, opening a market niche for Kubbernus. "I seem to have an appetite for doing things no one else wants to do. I've always found there are better opportunities that are more challenging and ultimately more rewarding when you go where no one has gone before."

In 1982, at age 21, Kubbernus put his growing business savvy to good use and founded his own company, Canyon HVAC – "I just liked the name" – and choose the image of a lone wolf on the horizon for the company logo, a nod to the remote regions work he was doing in the North West Territories. At the same time, he also bought/sold real estate and acting on his passion for motorcycles, he founded a Suzuki dealership that he later sold.

Concentrated on Niche Market

As a member of the Canadian Construction Association he was free to bid on various government projects and found there was less competition, fewer rival bidders and more likelihood of success when he went after remote area construction projects in the North West Territories such as RCMP stations or medical centres and schools. "I taught myself the bidding process and I continually tried to improve my bids for the greatest success" he recalls. "And I soon started winning my fair share of business at fair prices by taking on the toughest projects."

With the added prospect of reward came larger potential risks. "Mistakes can be costly. Northern shipping channels were only open for about 60 days each summer and any missed equipment or materials could mean financial ruin later. The last thing you wanted to do was rent a Hercules and build a runway in the middle of the frozen nowhere just to bring in the one piece of equipment needed to finish the job and get paid – while all the time paying rates of over $400 per night, per person in camp fees plus wages for crews waiting patiently for missing materials."

Then, in 1986, he sold his operation to a Winnipeg company, Landmark Mechanical, and stayed on another six months as a transition consultant.

At that point, Kubbernus decided to serve his other passion – for all things financial – and he took a year-long break to return to school, taking financial planning and business administration courses at Mt. Royal College and other post secondary schools.

Upon graduating, Kubbernus went to work for a financial services firm named Money Dynamics. "I mainly focused on business owners, executive asset management and strategic planning. By working closely with the owners and entrepreneurs I was inevitably dragged into the issues of their corporation and the financial health of their organizations. I soon learned that taking financial care of the executive always ended in taking care of the business."

In helping firms get on track financially, Kubbernus soon became known to the banking community. After he assisted a few small companies the banks began calling him on referrals to have him sort out bad loans and fix firms in trouble.

Money Dynamics was sold to PlanVest Financial (and was later sold to C.M. Oliver) and Kubbernus began working with the larger merged company, accepting numerous referrals and quickly establishing himself as the "resident turnaround guy."

In 1991 he founded Bankton Financial Corporation. Bankton's focus was to finance an increasing number of turnaround opportunities presenting themselves to Kubbernus as well as workouts, special situations and restructuring cases.

By this point in life the young entrepreneur had amassed a great deal of business savvy and expertise; discerning over the years what worked and what didn't in the world of business. "From travel agency chains to national 400 store clothing chains, we learned from our success, but we realized we learn even more from our failures and I knew that I could apply my knowledge to help others avoid making the same mistakes."

Founded Balaton Group

In 2000, Kubbernus moved to Toronto to get closer to a couple of large investments he made during the technology boom, including Jawz Technologies Inc. and Futurelink, both of which hit magnificent highs and equally magnificent lows with the tech melt-down of 2001. Later in 2004 he co-founded Balaton Group, also based in Toronto. In 2006, Kubbernus bought out his partners to head Balaton and build its market presence as a leading provider of turnarounds and restructuring measures to assist companies in trouble. Living in Toronto with his wife Becky and now three sons

Mason, Jackson and Christopher, Kubbernus finds home tends to be as interesting and vibrant as work.

Today, with several dozen successful turnarounds under his belt, Kubbernus acknowledges his expertise at turnarounds evolved over time. "I can't say I'm a gifted individual. The small and mid-size turn-around business is not something that can be taught as much as it is about experience. Other then some typical financial modelling courses I had to learn how to dissect financial statements to find the real problems. But many years later and countless projects behind me I now can get to the issues instantly and instinctually. I can smell it – I don't know where that comes from, except 20 years of experience and hundreds of crime scenes."

"Essentially," he adds, "we're corporate architects. Balaton Group, through tried and tested business practices, extensive worldwide networks and unsurpassed knowledge of business development, re-architect undervalued public entities to be prosperous industry leaders. Re-engineering a business involves identifying problem areas and capitalizing on existing strengths. To do this, a broad and comprehensive view must be applied. Often business professionals are mandated to evaluate and remedy underperforming companies. Far too often these undertakings are attempted with too narrow a view to be of any real value. At Balaton Group, we look at every single aspect of our business opportunities. No detail is too small, and no avenue of business operation goes unnoticed. We are determined to achieve our mission of creating long term value for our clients, investors, markets, and for ourselves."

Many Reasons for Undervalued Firms

Kubbernus notes companies can lose value via a number of factors, including real or perceive management failings, lack of focus or communication breakdowns.

"The companies we take on are undervalued for any number of reasons. We often take an ownership position in a company we're turning around, and once the firm is acquired, we remedy appropriate corporate market and capital issues; we reposition the business fundamentals, and then bring the company back into an industry-leadership position. We provide our portfolio companies with our significant depth and breadth of expertise on acquisitions, integration, strategy, negotiations, financing and public markets with a strong track record of completion. At Balaton Group, the "value shift" of a company begins with control, clarity and vision."

Kubbernus finds the process fascinating as he steps into varied companies, studies them and develops plans to turn around their failing fortunes.

"Every day is different – and there are so many companies out there that need help sorting out one problem or another. It's an interesting process in every case."

Non-obvious Cash Flow Problems

Kubbernus notes that in some cases, companies incur cash flow problems due to the desire of the managers to live a wealthy, successful life before the company is in a position to pay for it.

Such was the case with a homes and gardens magazine in Calgary. "It was a non-obvious problem," Kubbernus recalls, "because when I got there to conduct an analysis and sort out cash flow problems on behalf of its major investor, it was not apparent where these problems were coming from – the magazine was filled with large full colour display ads for expensive cars and stereos and other luxury items. The magazine was top quality and the company was running lean."

Kubbernus was brought in to analyze the perplexing situation of a seemingly successful magazine being $1.5 million in the red with serious cash flow problems.

"The company didn't have a chief financial officer so I became the interim CFO and went to work on the business plan and undertook a full analysis," Kubbernus recalls. "At first blush there did not appear to be any overspending on personnel. Freelancers did the writing and photography so those costs were fairly low. They were in inexpensive office space and the printers only wanted 50 per cent payments up front. The advertisers normally don't pay until they see the magazines finished tear sheets, but there were plenty of ads for Mercedes Benz and other luxury items so the advertising revenue should have been substantial."

As Kubbernus delved more deeply, it became apparent that there was relatively little advertising revenue. "It turns out that the partners were accepting 'contra' – swapping ad space for stereos and leased luxury cars – instead of charging money for a lot of the advertisements. "I needed to get invited to the executives' homes for dinner to get a full perspective on the situation. Just to make sure the benefits were not simply taken from salaries I had to cross reference all benefits against payroll. It turns out no deductions were made." The company was quickly restructured, luxury cars were returned and advertisers were told they had to pay for their ads as alternatives to cash would no longer be accepted.

Cable Channels to Social Networking

In some cases, the company name is kept confidential at the corporation's request. Such was the case with an online social networking company in 2002. This company was to become the first and only dating service to be messenger-centric. The Company's primary business was changed from Digital Cable programming to proprietary online/web and instant messaging technology hosting numerous social network opportunities.

Prior to enlisting the services of Kubbernus, the company was completely out of working capital; the stock price was approximately 4 cents making any fund raising activities doubtful; the firm had fallen out of favour with its shareholders; the original business plan had failed due to changes in the cable TV market and the technology sector was still suffering from the 2001 hangover of all hangovers. The company was in complete paralysis.

Kubbernus performed an initial review of the company and its circumstances at no cost. This review gave him enough information to determine what course to take.

The review included a complete analysis of the troubled firm's shareholder sentiment and capital structure, business plan, historic activities, management capabilities, contracts and board member capabilities. Lastly, an analysis of the company's technology was conducted as this was the foundation for the firm's future success.

Kubbernus found the company's business plan did not portray the expansive growth opportunities that the firms technology could achieve; very little of the company's research or knowledge base was visible.

The entire plan was re-cut using in-depth research on the global potential of the market and technology. After Kubbernus' team gathered all of the necessary raw materials, they re-wrote the plan.

Kubbernus advised that the look and feel of the firm, including its brand and image, had to radically change.

The institutional investor community cringed at any notion that the firm could be involved in pornography, which was embedded in the original business model, and this would ultimately prevent any direct or in-the-market investments in the company by institutions. "Institutions do not invest in Church's or X-Rated businesses. Both are equally problematic. The risk of being on the front page supporting either at the time of trouble is not worth any profit opportunities." Kubbernus laments. One of Toronto's best creative companies was hired to perform the corporate makeover.

Private Placements Utilized

With the new business plan and makeover in place it was now time to start working with potential investors for a series of private placements to facilitate the firm's business plan. To help ensure investors would receive the highest quality dissemination of information Kubbernus retained an investor relations firm to help put the company back on the radar screen, simultaneously raising capital and using his own internal communication group to give the story some visibility.

Work then began with previous investors, potential new investors and Kubbernus' investor network.

Each round of finance was significantly over-subscribed, which only helped stimulate each subsequent round.

Over a series of graduated rounds of finance, the firm had enough working capital to re-launch plus catch the attention of potential key executives to fill top roles in the company.

With a new business plan, a new brand emerging, working capital in the bank and investor relations working in lock-step,

Kubbernus was now ready to make sweeping management and board changes.

It is often very difficult for management changes to occur without the assistance of an impartial or outside group such as that which Kubbernus provides. "There is nothing sacrosanct when we step in. Everything is up for change."

But Kubbernus is guided by one basic principal: Meet and Exceed Shareholder Expectations. Although the stock price had been mired at 4 cents and a market capitalization of less than $1 million, the company was now trading at 60 to 70 cents and a market capitalization of approximately $24.5 million. Kubbernus received significant interest from a few parties who wanted to take control of the venture and continue with the growth of the company. They also committed to provide the next round of needed funding. Kubbernus turned over the reins after restoring the company to solid footing. From an original investment of only $750,000 handsome returns were made by the turn-around investment group.

Investors Rewarded Handsomely

Even more impressive investment returns were achieved in another turnaround, this one involving an oil and gas company (name withheld by request).

In the spring of 2003, Kubbernus stepped in to repair the public company which had lost confidence and its shares were trading at 7 cents.

The assets in the company were non-operating and the company was run by an absentee president put in place by the single debt holder over the company. The assets ranged from gold prospects to oil concessions in a foreign country.

Kubbernus brought in key investors from Switzerland and obtained control of the firm by paying out the debenture holder in full, recapitalizing the balance sheet and acquiring additional producing assets.

The early stage investors were rewarded handsomely: The Swiss initial investment of approximately $3.25 million doubled within 3 months and investors in the open markets saw the price run from 25 cents to over $1.30 within a 9 month period – and with solid volume.

Kubbernus achieved similar success with a German bio medical discovery company which commercializes patented and clinically tested natural products that relieve chronic pain.

The company was unable to commercialize its non-doping and highly developed pain relief products so Kubbernus was approached for assistance with business strategy, structure and finance.

Kubbernus, while working from Frankfurt, Germany, organized the corporate structure, arranged financing and concluded a trade sale into a public vehicle supported by a new management team located in North America.

The end result is a dynamic change in everyone's fortunes. The bio science and discovery products obtained finance, the share value grew from $.2 cents to $.70 cents and operations were moved to North America where the new management and a fresh board of directors took over.

Communications Company

Balaton Group is also expanding their line of services to go beyond its primary role as a leading corporate turnaround firm.

Kubbernus and company recently launched a new corporation under the name Balaton & Co. Inc, in response to the growing need for specialized corporate communications in the financial sector.

Balaton & Co. has several top tier clients that it has assisted with re-branding and communication campaigns with substantial success. Balaton & Co. boasts a team of design and communication industry experts that understand the small and mid cap space and the challenges these firms face.

Balaton & Co.'s team of experts delivers a full range of consulting and production services that helps their clients meet strategic marketing objectives.

"This is yet another exciting platform from which Balaton Group Inc. can leverage its talent and expertise to a whole new audience," notes Kubbernus.

"Creating a completely self-sufficient business entity that specializes in corporate communications for the small and mid cap markets and will significantly increase the chances of success for firms that occupy this space," he adds.

"Traditional large communication firms cannot respond as quickly with the right solutions, nor do they have the depth of specialized experience required to properly service the needs of this client base. As the micro cap and small cap turnaround industry's most respected authority on corporate re-structure, it's only natural Balaton Group Inc. would bring together a team of brand and communication specialists destined to become one of the most effective firms of its kind."

Turnarounds Remain Key Focus

Turnarounds will continue to be Kubbernus' main focus. One of the most recent turnarounds Kubbernus is orchestrating concerns an American firm – SkyPort – that delivers to very discreet customers – including the U.S. National Guard – a 100 per cent reliable communication connection services to ensure voice, data and other communications are successfully placed regardless of any external factors such as severe weather or war.

A combination of communications delivery vehicles – including satellites and fibre optics – are utilized and organized to ensure the system never fails.

The customers in turn pay a premium for secure access to a fail-safe communications system that also includes two levels of back-up power in case of power failure. Annual sales revenue had been ringing in at around $15 million US.

However, the four-year-old company was established by entrepreneurs and engineers who are not accustomed to describing their operations and cash needs in the precise manner the investment community feels most comfortable with, Kubbernus notes. "You can't just ask for a lot of money for product development without meeting certain financial reporting and accounting criteria and outlining in detail exactly how the money will be spent, the expected results and so on."

The firm's biggest investor, Century Tel, invested $20 million in SkyPort before deciding it would not back the fledgling firm any longer. The Century Tel board of directors felt the investment level of $20 million was sufficient and refused to entertain SkyPort's cash call requests for another $4 million. No other investors were interested in coming forward as they would be ranked secondary to Century Tel, a position unlikely to be acceptable to their shareholders. Kubbernus was called in to assess the situation.

"A restructuring of the company was the most appropriate approach as Century Tel did not want SkyPort to fail. We rejuvenated the management and reworked the business plan. We turned it around and bought the company at a significant discount. We have improved the balance sheet by just under $30 million, restored profitability and stabilized operations"

Cooperative Spirit Aided Turnaround

Kubbernus says the success of the SkyPort turnaround is owed in part to the entrepreneurial engineers who founded the company. He notes these executives could have sought and received golden parachutes that would have lined their pockets at the company's expense.

"Instead, the executives have shown an incredible amount of integrity and truly exemplary behaviour. Management has responded very well to the challenges and kept the emphasis and their focus squarely on helping the company succeed. To a large extent, they've also restored my faith in managers to act responsibly and helped me to further appreciate the value of these turnaround efforts in terms of saving companies and jobs and building sustainable prosperity. All too often the worst in people comes out during trouble times, however in the case of SkyPort the best surfaced."

Asked what continues to drive him to assist troubled companies, what motivates him to achieve greater returns for investors, Kubbernus says his core philosophy is best summed up in Balaton's mission statement: "Our objective is to create long term value for our clients, investors, markets, and for ourselves. Through management and vision, Balaton fosters company growth and development through capital market expertise and extensive business networking. We love what we do and we are the best at it and this is reflected in our portfolio companies' share prices."

Leadership Lessons:

1. Realize that there are no rules, just guidelines, and every situation is different; calling for tailor-made solutions.
2. Get rid of fear: If you think you can't succeed it's a self-fulfilling prophecy. Rise above fear and seize opportunity.
3. Challenge yourself to do more: The more tasks and challenges you take on, the better you become and the more you raise your threshold for pain and exhaustion. Simply put: The more times you take on the tough stuff, the better you get at it and the more you enjoy doing it.
4. Deal with the "tough stuff" – the difficult decisions – first and save the easy problems for last when you've got your business back on the right track.
5. Know when to reach out for help and don't be afraid to do so. Your own abilities and persistence can often take you most of the way but few people succeed entirely on their own. Help is often available when you most need it and it's amazing what you can get simply by asking for it.
6. Don't think that if you ignore a situation it'll eventually heal itself. It's more likely to get worse over time and become that much more difficult to solve as a result of your earlier inaction.
7. Don't hesitate on initiating a necessary decision or course of action simply because it's uncomfortable or distasteful. Delay simply adds to the difficulty in doing what must be done. Better to act fast, take the distasteful medicine, and get it over with. As Kubbernus asserts: "If you have to swallow a frog, it's best not to stare at it for too long."
8. Force yourself to solve problems and compare your situation with that of others and study how they resolved their difficulties. You'll be a better business person.
9. Always look for the unexpected whenever problems arise that seem to defy solution. Go over the details and find the missing pieces. The answer is in the problem itself.
10. Know your limitations in terms of ability. There's no shame in hiring experts to help – in fact it's a good idea.
11. Don't over-extend yourself – make sure your business is well financed and growing at a sustainable rate.

At a Glance:
CARSTAR Automotive Canada:

Sam Mercanti
Age: 59
Title: CEO and President, CARSTAR Automotive Canada.

Claim to fame: Among the many distinctions earned by Mercanti and his CARSTAR team are ISO 9000 status from the International Standards Organization; the Hamilton Chamber of Commerce Outstanding Business Achievement Award – large company category – for the year 1990; the 1997 Collision Industry Pride Award – Mercanti is the first Canadian to ever win the prestigious American award. He's also been twice nominated for Entrepreneur of The Year (1995 and 1997) by Canadian Business magazine; and he's won the Spirit of the Community Award from the Hamilton Safe Communities Coalition. As well, CARSTAR has been ranked among the top 200 fastest growing companies in Canada by Profit magazine.

CARSTAR: Canadian operations are part of the CARSTAR family of repair shops numbering more than 350 across North America – including 114 in Canada, employing over 1,200 people. Founded in the United States, the overall group of companies is the largest and leading collision repair group in North America.

Financial Data: Annual networked sales of Canadian operations alone exceed $125-million and the company is growing steadily.

Personal: Resides in the Stoney Creek Area of Hamilton with wife Roma. The couple have three daughters: Lisa, 34; Jennifer, 30; and Samantha, 25.

For More information:
Contact Sam Mercanti: (905) 388-4720. Fax: (905) 388-1124. Address: 1124 Rymal Road East, Hamilton, Ontario, L8W 3N7. Email: smercanti@carstar.ca

3

CARSTAR
Putting Business in the Driver's Seat

"A leader is someone who leads by example, who builds other leaders, who delegates authority and provides overall direction without micromanaging…"
- **Sam Mercanti,** President/CEO, CARSTAR Automotive Canada

Control over costs, assets and operations is crucial to any business, large or small. So what happens when a fender-bender damages a vehicle asset and strands a valuable executive? Do operations simply grind to a halt?

A growing number of businesses, and many consumers, are increasingly turning to CARSTAR, a leader in collision repairs.

CARSTAR locations across North America have set themselves apart from competitors by offering the industry's most comprehensive warranty. With stores in over 350 locations across United States and Canada, there's likely a CARSTAR nearby. If not, CARSTAR will find a convenient, quality, independent store to honour your warranty. And the warranty travels with you throughout the United States for up to five years, so your coverage continues even if you've moved or are traveling. CARSTAR offers a lifetime Nationwide Warranty within Canada.

The track record is also quite impressive: CARSTAR has repaired more than 2 million vehicles across North America. Every CARSTAR store is equipped with the latest systems and technology to get the job done right, right on time. They also pledge to keep you informed throughout the entire repair process. Not surprisingly, CARSTAR also leads the industry in customer satisfaction ratings.

Works closely with Insurers

CARSTAR also works directly and closely with most insurance providers to make the process as painless, fast and efficient as possible. It's likely that CARSTAR has a Direct Repair Program with your insurance provider, which will likely save time during the estimate process, as insurers have come to rely on the accuracy, validity and fairness of CARSTAR estimates and repairs.

As well, CARSTAR believes in having local owners run their stores. This means that when you visit your local CARSTAR, you'll deal with someone who is invested in the community, has put their own reputation on the line, and will strive to provide exceptional service.

In fact, the service is beyond exceptional: CARSTAR offers assistance virtually from the moment the collision takes place and they're contacted at 1-800-CARSTAR. They offer 24-hour roadside assistance and tow truck services, arrange rental vehicles, deal with the insurance firm, call friends or family members on your behalf, provide a Lifetime Nationwide Warranty, offer comprehensive claims assistance, perform high-quality repairs to all makes of vehicles to the manufacturers' standards – and even provide AIR MILES Reward Miles.

CARSTAR Automotive Canada

Sam Mercanti, CEO & President of CARSTAR Automotive Canada notes CARSTAR operations on both sides of the border pride themselves in being as helpful as possible.

"If you've had an accident, you don't need more stress in your life, so we're here to make the repair process as fast and painless as possible," Mercanti explains in an interview at CARSTAR Automotive Canada's head office in Hamilton, Ontario.

"We take pride in our guarantee that your vehicle will be repaired quickly, reliably and at a reasonable price," Mercanti adds with a smile.

Behind Mercanti, a picture window offers a view of Rymal Road – Highway 53 – teeming with cars – no doubt a number of former CARSTAR satisfied customers among them.

In front of Mercanti is a large desk with a Captain's Bell that's rung at media events each time a new franchisee joins the team. And there's the forward-looking gaze from a prominent bust of Napoleon Bonaparte.

"Napoleon was a great historical leader and an inspiration to me," Mercanti explains. "He was a great leader for what he was able to accomplish, coming from nowhere and overcoming tremendous obstacles to win many impressive victories and become the leader of a great empire."

Mercanti also approaches the financial side of the business with military precision, taking a numbers-oriented approach in keeping with the proclamation on his desktop plaque: "What gets measured gets done."

Measurable Goals

"I like attaching numbers to measurable goals," Mercanti asserts, "so I can determine if they've been achieved, and if not, how much more needs to be done. I measure everything – that was the basis for the first Ontario Auto Collision franchise and for CARSTAR."

Measured goals pursued by Mercanti also include stringent cost-containment efforts to lower average repair costs; and, decreasing the repair cycle time to "reduce the timeframe from when you give us your car keys to when you get them back."

Mercanti also monitors customer satisfaction: Each customer gets a car with a ratings survey in which 5 indicates the highest level of satisfaction. "If a card comes back with a rating lower than 4.6, the customer gets a call to help us determine what went wrong in their experience that put us below our normal high standard of customer satisfaction," explains Mercanti, who has been known to personally make follow-up calls to ensure customer satisfaction.

The diligent efforts are clearly bearing fruit: CARSTAR is now in all 10 Canadian provinces and boasts 114 repair shops nationwide. Annual sales across the network exceed $125-million. Brand awareness has risen to 16 per cent from 10 per cent of Canadians, helped somewhat by advertising on television's The Weather Network.

And the CARSTAR shops give back to their respective communities: In June 2006, 70 CARSTAR shops took part in National Car Wash Day and raised more than $60,000 in donations to support the Canadian Cystic Fibrosis Foundation. And local fundraising groups. Over the years, the CARSTAR repair centres have raised more than $1-million to support Cystic Fibrosis Research.

Finding new, inventive ways to lead

Mercanti notes CARSTAR "is committed to assisting our franchise partners in improving their businesses. At CARSTAR we are constantly working on new and inventive ways to lead, not follow the industry, and to follow our vision to be the leader in the Canadian collision repair business."

He adds that CARSTAR is also "driven by our vision: To build a chain of high-quality profitable collision stores across Canada operating with ethics, respect, integrity, standards and consistency. We promise customer satisfaction, cost containment, and improved turnaround time. The objectives are client retention for CARSTAR and our insurance partners, and profitability for all CARSTAR stakeholders. Our values include continuous improvement – we strive to do everything better – and recognition, by acknowledging the contributions of our people and of others."

One of the newest, largest and most advanced CARSTAR facilities is the 25,000-square-foot CARSTAR Collision Centre 401, just off Highway 401 in Mississauga. The impressive structure is on a 3.2-acre site on Argentia Road between Winston Churchill Boulevard and Erin Mills Parkway.

"It will deliver the highest level of cost containment, along with improved turnaround times and client retention for our insurance partners," Mercanti says of the sprawling $5-million facility, next to the Shell Service Centre.

"It's a new style of collision repair centre with state-of-the-art functionality, including modern technology and innovative and highly efficient management systems," Mercanti adds, noting the complex is located in Mississauga's fast-growing Meadowvale community, boasting a population of more than 200,000 people.

"And it's like the Home Depot of collision centres – it's the first retail driven collision centre in Canada where customers enjoy the ultimate collision repair experience," adds Mercanti, whose trademark slogan for the forerunner Ontario Auto Collision – "A bang-up job every time," – remains a familiar phrase for many Canadians.

Improving repair turnaround times

Mercanti says the industry average for collision repair turnaround – how long it takes for repairs to be completed and you get your vehicle back – is 11 days, while at CARSTAR, it's just 7 days.

Not satisfied with this level of performance, he wants to reduce the turnaround time to just under five days.

"We want to drive out inefficiencies and establish the best practices across Canada," he notes.

"We've hired process engineers to identify any inefficiencies in repairing a vehicle, from collision to completion of repairs. At this facility we'll do some teaching and demonstrate to franchisees how to adopt best practices," he continues.

"This is our starship location," he adds, "and it's where we'll develop new programs, new innovative ways to repair vehicles, faster, better, more efficiently and at less cost."

Mercanti notes CARSTAR 401 offers a number of benchmark services, including: 8 a.m. to 8 p.m. hours of multi-shift operation, seven days a week; Lifetime Nationwide Warranties; VIP pickup and delivery shuttle services; on-site car rental agency; interior and exterior detailing with every repair; and 24-hour accident assistance and towing services.

"We're incorporating quality control processes to reduce defects to less than 1 per cent. And we'll offer all of this using the new, highly efficient CARSTAR Operating System to manage information and data bases, which in turn should enable the insurers to be more profitable through better risk management," adds Mercanti, whose vision has guided this business through each stage of its corporate evolution. His hands-on management style continues to drive CARSTAR to new levels of success.

Mercanti Credits Others

Mercanti credits CARSTAR's impressive success to the combined efforts of the various location managers; suppliers; insurance partners; CARSTAR staff; and his founding partners: uncles Nardino, Guerino and Anthony Mercanti; along with a talented executive, which includes Executive Vice-president and co-founder Larry Jeffries and Chief Financial Officer Paul Tice. In May of 2006, the team expanded to include Vice-president of operations David Lush.

Of course Mercanti's own driving vision should not be overlooked – and the repair industry has certainly taken notice: Among the many distinctions earned by Mercanti and his CARSTAR team are ISO 9000 status from the International Standards Organization; the Hamilton Chamber of Commerce Outstanding Business Achievement Award – large company category – for the year 1990; the 1997 Collision Industry Pride Award – Mercanti is the first Canadian to ever win the prestigious American award.

He's also been twice nominated for Entrepreneur of The Year (1995 and 1997) by Canadian Business magazine; and he's won the Spirit Of The Community Award from the Hamilton Safe Communities Coalition. As well, CARSTAR has been ranked among the top 200 fastest growing companies in Canada by Profit magazine.

Active Community Supporter

Mercanti has long been active in assisting the community, helping many worthy causes.

Beyond his support for the Cystic Fibrosis and the Safe Communities Coalition, he sits on the board of trustees for St. Joseph's Hospital in Hamilton.

He's also spent more than a decade as the chairman of the advisory committee for the Centre For Ambulatory Health Services in Stoney Creek.

As well, Mercanti heads a Christian men's self-improvement group aimed at creating better husbands, fathers and businessmen.

How CARSTAR Automotive Canada and Sam Mercanti achieved such enviable success is a story that has its beginnings in a small town in Italy.

Born In Italy

Sam Mercanti, now 59, was born Sept. 14, 1947, to Giuseppe and Iolanda Mercanti in the scenic village of Castelli, in the mountainous Abruzzo region of north-central Italy.

He was the eldest of four children and would soon be joined by his brothers Peter, now 57, and Morris, now 53, while their sister Rosanna, now 43, would be born in Canada.

The picturesque village is situated on a rocky plateau in the shadow of snow-capped mountains.

Approximately a two-hour drive from Rome, the community is famous for its Castellian ceramics, drawing throngs of tourists eager to buy hand-made ceramic pottery, tiles and works of art and travel through the scenic surrounding farmlands.

Despite the idyllic setting, the lack of employment and career opportunity made life less than ideal for the young family in the quaint community of less than 2,000 people.

Parents Sharecroppers

"My parents were sharecroppers, farmers who didn't own any land and it wasn't easy for them to provide for a whole family on what they received from their work," Mercanti explains.

"The family's ties went back forever in the little town and I can remember going to school there and how hilly the area was," he continues.

"I can also remember drawing water from a mountain cave – and when I went back for a visit, decades later, I found the farm was abandoned and the cave opening grown-over, but once I cleared the bramble there was the well and the water was still cold and refreshing."

"My parents worked on a hillside farm, about the size of a hobby farm. It was very hard to make a living. There was little opportunity for success in the community, so by the time I was nine-years-old, my parents had decided to move to Canada and Hamilton."

The Move To Canada

In 1956, the Mercanti family immigrated to Canada, arriving by ship at Pier 21 in Halifax. Then came the long train ride from Halifax to Hamilton's CN rail station at James Street North.

"My family was given two loaves of white bread during the train ride, and we thought it was cake – we never put sugar in bread in Italy, only salt," Mercanti recalls.

Despite other such adjustments to life in Canada, Mercanti has no regrets. Far from it.

"I thank God my parents decided to take us all to Canada, to take that risk, to leave their home of many generations for a new land of opportunity," Mercanti states, his voice catching in his throat.

"And I'm especially grateful they came to Hamilton for the opportunities here. Some people say I've got an enterprising spirit – but that never existed until I came here. My parents gave us the opportunity to succeed. They did it all for their children."

His mother picked fruit on area farms while his father initially worked as a construction labourer.

Then, in 1957, his father went to work for his third cousins, Frank and Ralph Mercanti, who had started the first Mercanti autobody business in 1953 in the Bay and Vine streets area.

Half-Century Tradition

The Mercanti name has now been ingrained in this business for more than half a century.

"My dad usually prepped cars for paint jobs," Mercanti recalls, "and in the summer time, Dad would take me to work with him. He taught me how to prep cars and do detail work when I was about 15 years old."

That was Sam Mercanti's introduction to the auto restoration business. And at the same time, his entrepreneurial streak began to emerge.

Enterprising Youth

While still in their teens, Mercanti and his brothers were involved in several enterprising, entrepreneurial ventures, including some in which timing was everything.

"My brothers and I would buy 20 copies of the Saturday afternoon edition of *The Hamilton Spectator* at 6 cents each and then run to the street corners and sell them for the full regular price of 10 cents a paper," he recalls with a smile.

"A lot of kids did the same thing – and if somebody took somebody else's corner there'd be fist-fights – it was very competitive."

But the Mercanti boys entrepreneurial drive didn't end with the quick profit derived from selling newspapers at nearly twice their bulk-purchased price.

That would have been impressive enough – but there's more to this enterprising story:

Once the papers were sold, the brothers were off and running again, this time to the Hamilton Farmer's Market where they rushed up to a florist selling fresh-cut roses.

"The roses were cut fresh earlier in the day," Mercanti points out, "so by the time we got there, any unsold roses would soon be in danger of wilting."

Always willing to help out a florist in need, the Mercanti brothers struck a deal.

"We'd buy up their fresh-cut flowers for just 25 cents a bunch," Mercanti grins, "and then we'd stand outside the downtown hotels and sell them to the ladies' escorts for $1 a bunch.

Faced with an enterprising young man selling roses, a gentleman's only correct response was: Yes, he'd be delighted to buy roses for his lady friend.

"And we got tips," Mercanti laughs, shaking his head. "We'd head home Saturday night with $50 between the three of us. It was outrageous."

Valuable Lessons Learned

Mercanti learned some valuable lessons in the process.

"We learned to buy low, sell high and be quick about it. And timing is everything: A day-old paper is worth nothing. Dead flowers are worth nothing. You have to hustle and move fast to seize opportunity."

And Mercanti's unquestionable enterprising spirit continued undiminished into high school.

While attending Westdale High, and then Central High schools, Mercanti became aware of another consumer need waiting to be filled.

"There weren't many dance halls in the area, especially for young people," he recalls. "So I decided to run my own dance hall. I talked to the owner of a pool room on James Street North at Wilson Street. The upstairs room was just sitting there empty, so he let me fix it up and use it as a dance hall and we split the earnings. We had great turn-outs. It was a lot of fun – and very profitable too."

Joins Auto Body Shop Business

In 1964, Mercanti, then 17-years-old and eager to continue seeking opportunities in the working world, dropped out of high school in Grade 10 and went to work for Mercanti Brothers Auto Body run by his uncles Nardino and Guerino Mercanti on Strachan Street.

"I started doing detail work and prepping cars for paint jobs," he recollects. "Then I got my technical body man's licence and I fixed cars."

In 1967, Mercanti found himself writing estimates. His uncles had some lingering difficulty reading and writing in English and the young nephew – then barely 20 – enjoyed working the front desk, writing estimates and dealing with customers.

The Mercanti brothers' auto body shop property was expropriated by the City of Hamilton in 1968 for civic development so they moved their business to a new location, opening shop on Gage Avenue North under the new name: Ontario Auto Collision.

At that time, a third uncle, Anthony Mercanti, came into the business and the growing shop needed a general manager. The uncles turned to their hard-working nephew, the one with the entrepreneurial spirit, the gift of the gab, the sales and marketing skills and the urge to run things. Sam Mercanti was moving up in the world.

In 1968, Ontario Auto Collision was then averaging $100 for some body work and a paint job on used vehicles.

At that time, the bulk of the company's business was from used car and warranty jobs and only 5 per cent from insurance companies.

Insurance Claims Became Focus

"That same year," Mercanti says raising an eyebrow, "we had an insurance job come in – a 1966 black Pontiac – and this two-year-old car needed some work. It was a $1,100 job – eleven times what we were averaging with most jobs. It did need a little more body work and paint and parts. But we'd have to paint 11 cars to get that kind of money. And the insurance jobs are very stable – you're definitely going to get paid – and there's a better gross profit."

For Mercanti, the insurance job experience was nothing less than a revelation.

"I decided to go after the untapped insurance business and I met with brokers, agents and appraisers to assess their needs and earn their business," recalls Mercanti, who also succeeded in raising Ontario Auto Collision's consumer market presence with his famous marketing campaign and TV commercials promising "A bang-up job every time."

"The insurance industry was key," he adds. "I realized we needed to put our focus there, so I put together a sales force and we became the first auto body shop to have a sales force dedicated to bringing in business from insurance companies, insurance brokers and owners of fleets of vehicles."

Although Ontario Auto Collision continued to serve a wide clientele including people from all walks of life, it concentrated on bringing in insurance business.

The sharpened focus soon paid off: In 1968, Ontario Auto Collision was a $100,000-a-year business.

In 1972, just four years later, it had grown to become a $1-million-a-year business.

And insurance jobs accounted for 50 per cent of revenue. This explains why Mercanti is one of the few people who speak of insurance firms with genuine affection, and in reverential tones usually reserved for sports heroes.

"A lot of thanks has to go to the insurance companies, they embraced me early on and the business relationship is very good," Mercanti asserts.

An Important Year

For Mercanti, 1972 was a pivotal year in more ways than one.

"I married Roma, the girl of my dreams," he says grinning widely. The couple have three daughters: Lisa, 34; Jennifer, 30; and Samantha, 25.

Also in 1972, with the company doing nearly a $1-million annually, Mercanti was getting restless and was contemplating opening up his own shop and going into business for himself.

After making his uncles aware of his thoughts, they realized there was only one thing to do: They made their nephew president of Ontario Auto Collision and a full partner with 25 per cent ownership of the business.

In 1978, Ontario Auto Collision crossed the $2-million-a-year revenue threshold and needed more room to accommodate its fleet of 10 vehicles and burgeoning business.

The company soon acquired neighbouring buildings and land for parking and expanded to include 10 service bays, a larger office and waiting room, and additional customer parking.

Next, the Bank of Montreal loaned Ontario Auto Collision its first $200,000 business development loan and the company built a 10,000-square-foot, state-of-the-art facility.

In 1982, the company was doing $3-million in annual sales and was developing a strong management team that came to include his cousin, Tony Mercanti, son of Guerino.

Having establishing itself as a multi-million-dollar company, Ontario Auto Collision acquired the former McPetrie Motors Company in Burlington – the land, building, business and assets – and Mercanti took over its operations as manager.

In just two years the newly acquired business went from $400,000 in annual revenue to more than $1-million.

In 1987, Mercanti established Ontario Trucking Division for collision repairs to trucks and OAC opened its first Ontario Trucking Division store at a Gage Avenue location.

The same year, Mercanti turned over the management of the Burlington location – renamed as Ontario Auto Collision – to another cousin, Dino Mercanti, the son of Nardino.

Also that same year, Mercanti established OAM (Ontario Auto Management) and he founded the property management arm GNAS (named for the partners: Guerino, Nardino, Anthony and Sam).

And, still in 1987, the company did another acquisition, buying the land and building for its Ontario Auto Collision location on Highway 53 in Ancaster. Sam Saputo, the production manager at the Burlington store, was transferred to the Ancaster store as general manager.

Headquarters Built

The following year came the 1988 acquisition of land on Highway 53 near Upper Ottawa Street on Hamilton Mountain. And in 1990, Ontario Auto Collision built its signature $5-million, 36,000-square-foot building housing its headquarters, an auto mall, body shop and 15,000 square feet of leased-out units.

Another cousin, Remo Mercanti, son of Guerino, was named corporate locations manager, while his brother Sam, another Sam Mercanti, managed one of the OAC locations.

In 1992, OAF (Ontario Auto Franchise) opened a location in Stoney Creek and put Domenic and Rosanna Lucarelli at the helm.

Lucarelli? "We were running out of Mercantis to manage locations," Mercanti shrugs with just the hint of a smile. "Domenic was the production manager at our head office site and he had plans to open either a body shop of his own or a catering business. I didn't want to lose this guy – he's a great entrepreneur and a hard-worker – and I know how it is to be restless, so I helped him start-up the Stoney Creek franchise."

In 1993, The Ancaster location was franchised and was sold to managers Sam and Rosa Saputo.

The year 1993 marked the 10th anniversary of Mercanti's 1983 founding of a management system: CARS (Computerized Automotive Repairing System). It had provided a new organizational dimension to the business with consistent standards and business ethics at all locations.

In 1994, Ontario Auto Collision franchised its Burlington operation. OAC was by now a highly successful franchising company with eight locations across Ontario and $12-million in annual sales.

As well, Mercanti had by now made Ontario Auto Collision a household name.

Industry Changing

Still, he detected dark clouds on the horizon.

"I saw the industry was changing, and usually when that happens, regional players get clobbered."

Mercanti looked south to the fast-growing CARSTAR auto collision repairs franchise company in the United States, which then boasted 360 locations in the U.S. After carefully examining the pros and cons of an alliance, Mercanti signed an agreement and acquired a master licensee agreement for all of Canada in 1995. He now had the exclusive right to open and license CARSTAR locations across Canada. This marked the start of CARSTAR Automotive Canada, and the beginning of many good things to come.

Next, Mercanti assembled a team, including Larry Jeffries from the giant BASF paint company, as vice-president of operations, to create a network of CARSTAR collision stores across Canada.

"A strong management team is essential for sustained success," says Mercanti, whose privately held company is 82 per cent owned by the Mercanti family and 18 per cent owned by the management and staff at CARSTAR Automotive Canada.

High Common Standards Set

Mercanti says the core strategy was to ensure all of the stores awarded franchises met the same "ethics, standards and consistency that will deliver cost containment, improved turnaround times and customer retention for our insurance partners and CARSTAR stakeholders – that was and is our vision."

That same year, Mercanti began converting existing and acquired franchise operations to make them CARSTAR Automotive Canada locations. The company also began acquiring new locations. It bought two stores in Winnipeg, one in Calgary, one in Montreal and three in the Greater Toronto Area.

That brought the number of locations to 15. Two years later, in 1997, CARSTAR sites numbered 50 and annual sales more than doubled to $25-million.

Insurance Companies Partners

"The insurance companies are really our insurance partners – we work closely together," notes Mercanti, who has succeeded in earning a growing share of the auto body repairs market. "Our goal is to have $300-million in annual sales and 200 locations across Canada over the next few years."

"Shorter-term goals," he adds, "are to invest more than $1-million on information to connect all our stores by a single computer network and to improve our connectivity to insurance partners and brokers. We're also bringing in a pension/retirement plan for our employees and improving our recruitment of talented technical and management people."

For a potential franchisee the benefits are many, including advertising and promotional exposure and the cache that comes with a national brand.

Automatic Referrals

But the big benefit is the company's sterling reputation, which virtually guarantees business.

"Once we approve a franchisee, the insurance companies will automatically refer accident claims taking place in that area to that franchisee," Mercanti explains, "because the insurance companies tend to refer their clients to the nearest CARSTAR location."

The faith insurers have in CARSTAR is the result of many years of exemplary service and dedication delivered by Mercanti, who has still further increased the trust factor by using digital camera technology in the collision centres to provide computer-driven video imaging, allowing insurance appraisers to get a good look at vehicle damage without have to take the time to leave their office.

Repairs Faster, Better

The result: Repairs are authorized and completed faster, with better quality than the average body shop – and with less inconvenience to consumers.

"Based on our reputation for honesty and integrity, insurance companies will automatically approve our repair claims in most instances without involving an appraiser," notes Mercanti. "And that saves everyone time and money."

"We're a customer focused business," he continues, "and that includes national guarantees – recognized at CARSTAR stores across Canada and lifetime warrantees on our work. We want to do more than just satisfy the customer – we want to go the extra mile and delight them."

It's all part of an accelerated evolution that Mercanti envisions for CARSTAR and the industry.

"We're trying to elevate the body shop industry to higher standards," adds Mercanti, whose company is highly ranked by the Coyote Vision Group, which represents the North American industry.

Mercanti is marketing his business system to the franchisees with customer satisfaction at the heart of a strategy to make CARSTAR as just familiar to average consumers as Speedy mufflers and Harvey's hamburgers.

"Our objective is to establish a brand presence in Canada," Mercanti explains. "Right now, there's no nationally recognized collision centre brand presence. We want to change that, so if you get in an accident and your insurance company asks you where you want to have your vehicle repaired; CARSTAR comes to mind right away."

Growing Market Share

That key objective is becoming a reality: When CARSTAR Automotive Canada was founded in 1995; it had 0.5 per cent of the Canadian collision repairs market. By the end of 2000, market share had risen to 4.2 per cent.

By 2003 it had surpassed 5 per cent – and had reached levels as high as 18 per cent in Hamilton-Niagara and 14.7 per cent in the Greater Toronto Area where CARSTAR locations are most heavily concentrated.

Total national market share was estimated at around 6.5 per cent in 2006, with 7 per cent share expected later in the decade.

Brand awareness has also been rising steadily from zero in 1995 to an estimated 17 per cent by 2005, meaning nearly one person in five is likely to name CARSTAR as the company name that comes to mind when asked about their preferred repair location following an accident.

While remaining sharply focused on the future, Mercanti takes a moment to reflect on values and life lessons that have taken him to this point.

"My parents have been a huge influence in teaching me values," Mercanti acknowledges. "They're very humble people but also very proud and willing to do so much for their children. Family means a lot. My parents came here with no money, not even knowing how to speak the language. But they had family: They had nothing and they had everything. They gave me great values that have stayed with me my whole life."

Mercanti says success is never achieved alone or in isolation. It's owed to relationships with others.

It's All About People

"When you get right down to it, success in life is really all about people, and building relationships. My wife Roma, our children, our family – that's the focus of my own life, why I do what I do," asserts Mercanti who treasures family time. The longtime YMCA member is also an ardent Handball player.

Treat Others as You'd like to be Treated

"I learned from my parents to treat others as I would want to be treated to approach everything with ethics and honesty, to earn the trust of others. For our customers, it's all about efficient, quality work at a reasonable cost, based on trust and dependable performance. Walking your talk – that's so important.

Mercanti intends to continue this approach to life and achieving lasting success.

"We're building our success one relationship at a time. And we owe our success to our management, our stakeholders, our franchisees, our insurance partners, our vendors, our customers and this incredible country."

Leadership Lessons:

1. A good leader leads by example, builds and empowers other leaders; provides direction without micromanaging.

2. Praise employees for what they do right; accept that they make mistakes and help them correct and learn from them.

3. Not everyone wants to lead: Do not force leadership on anyone, but give opportunity to those who do wish to lead. Groom the next generation of leaders for challenges ahead.

4. Get in touch with customers: Get out on the shop floor or the street to get a feel for how your company is perceived.

5. Surround yourself with talented people and draw on their expertise to win business battles.

6. Adopt strong values and ethics and live up to them.

7. Establish and nurture relationships with other people. Treat others as you would like to be treated.

8. Don't die with your dreams still in your heart – make your dreams a reality.

9. Try to be all that you can be. It means hard work but the results are well worth it.

10. Don't deny yourself the pleasures of time with loved ones.

11. Take calculated risks; there is no reward without risk.

12. Go out and achieve – give life your best shot.

13. Internalize a belief system that has confidence in your ability to conceive, believe and achieve.

14. Don't just satisfy your customers. Delight them by going the extra mile and delivering added value.

15. Believe in yourself. We are capable of doing whatever we put our minds to. Grow each day in mind, body and spirit.

"...books have an almost magical aura about them; they command and receive respect instantly; they're imbued with high credibility and believability in a manner newspapers or other publications rarely achieve; they make us want to spend time with them and share the thoughts and messages in their pages; they're master communicators..."

"Writing a book – or being featured in one, imbues you with enhanced credibility and sets you apart from competitors who lack this connection. A book differentiates you from others in your field; it establishes you as an authority, a recognized expert; it states you're someone special with something to say of importance and meaning; it says you've arrived..."

"Manor House Publishing is uniquely positioned to provide a start-to-finish experience in creating and releasing superb business books to a global audience. At the start, we revamp concept, structure and content to improve the book's marketability. Throughout, we provide expert writing, editing, layout and design work. Then we complete the project by publishing, promoting and distributing it to worldwide markets."

- Michael B Davie, President, Manor House Publishing Inc.

4

Manor House Publishing
Sharing Your Business Message with the World

At a Glance:
Manor House Publishing:

Michael B. Davie

Title: President, CEO, Founder of Manor House Publishing Inc.

Claim to fame: The creation of Ancaster, Ontario-based Manor House Publishing came about from a desire to take the creation and publishing of books – particularly business books – to a higher level. Manor House works closely with talented entrepreneurs to help them achieve greater success through the creation, publishing and release of superb business books that relate their key messages in a clear, effective manner. Davie was instrumental in creating, publishing and marketing of such innovative business books as *Success Stories*, *Enterprise 2000*, the *Winning Ways* series, the *StreetSmart Marketer* series, and, of course, *Leaders*.

Manor House Process: In most cases, the business executive or entrepreneur lacks the time and some of the skill required to write a book. Manor House retains the author's role as writer, but assists in providing structure, marketable theme and format. Professional writers, editors, proofreaders and design experts assist the author in the creation of the book, working closely *with* the author's writing – not replacing it – to help craft a polished finished product that expresses key messages clearly and effectively. The book is then published, placed in bookstores and marketed globally.

Financial Data: Privately held company achieving 50-100 per cent annual revenue growth rate.

For More information:
Contact: Michael B. Davie: 905-648-2193 FAX: 905-648-8369
Email: mbdavie@manor-house.biz
Address: 452 Cottingham Cres., Ancaster, ON, Canada L9G 3V6
Website: www.manor-house.biz

4

Manor House Publishing
Sharing Your Business Message with the World

Being an author or featured subject of a book is often proof you've made the big time; you're a recognized expert. Books, after all, convey effective, lasting, high-credibility messages.

But many business leaders find dreams of sharing ideas and advice in a book can become a frustrating nightmare.

More often than not, their book is never published. Not only the entrepreneurs lose in such situations – the reading public also loses, having been denied the opportunity to read books that could have helped them solve problems, achieve greater success; change their lives for the better.

While a fortunate few do get published, most amass a stack of rejection letters from agents and publishers alike. In many cases, the rejections are owed solely to the fact the author is a relative unknown. The message, concept and quality of writing are not considered. Nor does anyone take the time to suggest changes to the manuscript that might broaden the appeal, potentially transforming a small-market book into a bestseller.

Some business leaders take a slightly different ill fated approach: A staffer, who may have once written a press release, is assigned the task of writing a book; a task much easier said than done. The project either falls short of expectations or fails entirely.

In other cases the business leader hires professional writers – who may or may not be qualified – and brings in costly editors, only to find the finished manuscript is still rejected for publication.

Some decide to go the costly self-publishing route, paying big sums to have the book prepared and printed, but with little chance bookstores will carry it, as most big bookstores will not carry self-published books or vanity press books or books that are printed digitally instead of higher quality web-offset press.

In many cases the self published book is poorly designed and laid out. The cover is drab or garish with a self-published look few will take seriously. Worse, the ugly end product may harm rather than enhance the business leader's reputation.

The Manor House Publishing Approach

This is where Manor House Publishing Inc. can help with a less costly, highly effective professional approach that focuses on the author and/or client business and keeps them involved every step of the way to ensure the end product meets or exceeds expectations. The book is then published and marketed worldwide, giving authors or participants an enhanced international image.

Michael B. Davie, president of Manor House Publishing, takes an innovative and thoughtful approach that often means thinking outside the box to create exceptional books.

For example, while many publishers rarely provide or seek input, Davie takes a far more proactive approach. "We always make a point of discussing the project in detail, determining what the author or business is looking for, and then finding ways to broaden the market appeal to increase sales to everyone's benefit."

"Sometimes," he adds, "it's as simple as changing the book's title or focus to something more effective or adding in some additional content – but these things can make a huge difference in a book's success and assist our worldwide marketing efforts."

Some recent examples of this approach include: revamping the new Michael Hepworth marketing book, including shortening the title to *The StreetSmart Marketer* and changing the format to include an engaging storyline coupled with Hepworth's terrific marketing advice, changes that should help this soon-to-be-bestseller gain even greater recognition and sell even more copies.

And of course, there's this book, *Leaders*. Davie came up with the book's title but wanted to go beyond the initial concept of simply profiling successful business people who are leaders in their respective fields. His concept: narrow the focus to those successful businesses that help other businesses succeed, and in so doing, create an original book that attracts a far wider audience. He also contributed extensively to the writing, editing and structure of the book, assisting with interviews to get to the heart and soul of the featured companies. I learned a great deal from this process and know that it strengthened my book immeasurably. He also revamped the secondary title make the focus clear: *How Top Innovators Can Help Your Business Succeed On a Global Basis.*

A Superior Option to Self-publishing

Although he understands why business leaders turn to self publishing in frustration, Davie cautions against anyone dealing with vanity press, self-publishing shops and associated operators.

"On the one hand," he notes, "there's the small 'plus' that these operators can drive added business for our publishing house – because they make us look very good by comparison. That's owed to the fact that just about anyone who has used a vanity press and paid to have their book published is usually unsatisfied with the results and has paid a fortune to have a basement full of books they can't sell. When they then come to us, they have an idea of just how difficult and costly publishing can be. They're extremely appreciative of anything we can do for them and are eager to promote, sell and/or purchase books, anything to make their project succeed. That's in contrast to some other first-time writers with no understanding of publishing and unrealistic expectations."

"However," Davie quickly adds, "that small plus is greatly outweighed by the negatives; the fact that this added appreciation of the publishing business has often come about through the school of hard knocks. Many of the people who have encountered the vanity press/self publishing experience are, in my view, victims of operators who take their money but give little value in return."

In one case, Davie recalls a distraught writer came to him with news that a "professional editor" had deemed her work to be childish rubbish – but for a fee of $20,000 he'd bring it up to publishable standards. She would then be on her own to find a publisher. "We decided to publish her book. We edited, revamped, designed and published it, at no cost to her. We sold 1,000 copies around the world and paid her thousands of dollars in royalties."

Sometimes, however, the self-publishing illusion of control can be powerful. "One business woman approached us to publish her book and was very interested in purchasing large quantities of books to resell on her website," Davie recalls.

"But she balked on being told we'd provide a 40 per cent discount on her purchases – the same rate given to bookstores. She said a vanity press would give her a bigger discount if she self-published with them. I asked what her costs were and she said for $10,000 they'd publish her book, provide two 'free' copies and give her the right to buy more at 90 per cent off $49.95 list price."

Sounds like a good deal? Or was it? Davie explains: "The first thing I told her was that since we'd publish at no cost to her, she was effectively buying the first two copies of her book from the vanity press at $5,000 per copy before getting a lower rate. I also said few will pay $50 for her book as it's double what the retail rate should be. I then told her if she bought 100 more copies she'd have paid $5,400 for 102 copies – a blended rate of about $53 for each copy of a book that will only fetch $25. Unfortunately she thought the vanity press route offered her more control – an illusion of course – so she went with them. The last I heard; she had a basement full of unsold books. You can't save everyone."

Worldwide Market Presence

Davie asserts: "Manor House was born in part from a desire to offer a viable alternative to two bad options: traditional publishing houses that instinctively say 'no' without ever trying to make a book work; and those self-publishing vanity presses that gouge customers with high fees to slap ugly books together with no effort made to market them or get them placed in bookstores."

In addition to providing editorial expertise and structural and format consulting, meticulous editing and rewriting services, Manor House also endeavours to get the book placed in bookstores, at times securing front shelf New Releases placements. Manor House distributes across Canada and works closely with its international distributors based in the U.S. and U.K. to give the book a strong presence in markets throughout the world.

The positive impact can be astonishingly dramatic for any author or person featured as the subject of a professionally presented, well edited, well structured and engaging book with an attention-getting cover – all of which Manor House routinely provides. Such a book can suddenly catapult them into a new stratosphere of respect and recognition on an international level.

The High-Credibility Magic of Books

"There's no question," Davie observes, "that books have an almost magical aura about them; they command and receive respect instantly; they're imbued with high credibility and believability in a manner newspapers or other publications rarely achieve; they make us want to spend time with them and share the thoughts and messages in their pages; they're master communicators."

"Writing a book, – or being featured in one," he continues, "imbues you with enhanced credibility and sets you apart from competitors who lack this connection. A book differentiates you from others in your field; it establishes you as an authority, a recognized expert; it states you're someone special with something to say of importance and meaning; it says you've arrived."

"Manor House Publishing is uniquely positioned to provide a start-to-finish experience in creating and releasing superb business books to a global audience. At the start, we revamp concept, structure and content to improve the book's marketability. Throughout, we provide expert writing, editing, layout and design work. Then we complete the project by publishing, promoting and distributing it to worldwide markets."

Davie notes that while some business books are aimed at global markets, others are aimed purely at niche markets and some are intended mainly for a select family of businesses as something of an in-house affair such as a particular firm and its clients and suppliers. "We can work with any size of market and happily consider every project proposal that comes our way."

"I should mention though that we receive submissions from all over the world on a daily basis and cannot possibly publish everything," he cautions. "That said, however, we do examine everything that comes our way and often find mutually rewarding opportunities in projects other publishers have rejected or ignored. We tend to see potential where others do not. We have the ability to perceive how a project can succeed with key modifications."

Ancaster, Ontario-based Manor House Publishing has grown steadily since its founding by Davie in 1998 and in recent years has enjoyed annual revenue growth rates of 50-100 per cent.

This heady growth in revenue is owed in part to the increasing ability of Manor House to tap into markets and steadily expand sales revenue.

How Davie came to be at the helm of a rapidly growing publishing house serving international markets is an inspiring story that begins in Canada's steel city, Hamilton, Ontario.

The Manor House Story

Born in Hamilton in 1954, Davie developed in interest in writing at an early age and was contributing articles to school publication in his mid-teens in the late 1960s, while working on his first novel, *The Late Man*.

In the early '70s he was contributing articles to alternative culture publications and in the mid-1970s he turned professional as the paid editor of The Phoenix serving the students of Mohawk College where he took journalism (he has an extensive education, Honours degrees in Political Science, and genius-level IQ).

On graduating, Davie spent about five years with various Canadian newspapers, then joined The Hamilton Spectator in 1980, where he won 28 awards for outstanding journalism during 17 years in a range of roles, including reporter/editor, Queen's Park Bureau Chief, feature writer, and senior City Hall reporter. His stay also included 10 years as a Business section writer.

While with The Spectator, his reputation for excellence as a business writer met with an invitation in 1995 to write business profiles for a coffee table book commemorating Hamilton's sesquicentennial in 1996. This role was expanded by the publisher, BRaSH Publishing, to include a chapter in the book and Davie then joined Sherry Sleightholm as a co-author of the book. The experience left a lasting impression on Davie.

"This was the first book I'd ever had a role in writing – and I found it to be a very rewarding and satisfying experience," Davie recalls. "Books last forever – they're immortal – and I loved being part of the creative process."

But Davie also envisioned an entirely new concept for the standard community-focused coffee table book: "It occurred to me that the business profiles were being scattered throughout the *Hamilton* book with little reasoning behind their placements. I'd written all of them as interesting pieces readers would find enjoyable. I came up with the idea of a hybrid book, a combination business book and community-focused coffee table book in which the business profiles would be a core and integral part of the book and the community aspects would largely focus on the economy and wider business community. The profiles would be expanded into short chapters and people would pick up such a book primarily because they wanted to read the stories of how prominent business leaders achieved their success – and I'd call it *Success Stories*."

Success Stories

With the release of the *Hamilton* book in 1996, Davie ran his *Success Stories* concept by BRaSH publisher Bruno Ruberto, who readily offered to publish it in project-partnership with Davie.

The timing could not have been better: The Spectator was offering employees a voluntary severance package with several weeks pay for each year of service, for any employee interested. Davie confirmed he'd take the VSP, realizing it would serve as a financial cushion while he gathered information and wrote the book. It would also make the transition to book writing and publishing easier by removing some immediate financial risks.

Released in 1997, *Success Stories* was filled with captivating profiles of some of the most successful business leaders in Canada. Davie interviewed Tim Hortons CEO Ron Joyce at his private airport, and flew down to Bermuda for a week to interview another billionaire, Michael DeGroote, who built Laidlaw Corp. into a multi-billion-dollar behemoth before starting a new business empire in paradise. Davie's photographer wife Philippa took stunning photos of the billionaire businessman aboard his yacht and amid tropical splendour. The book was winning combination of great stories and photos and was a huge financial success.

Realizing that the market would not support such expensive books on an annual basis, Davie accepted an editor position with Canada's national newspaper, The Globe and Mail while he worked on a sequel to his *Success Stories* book to be released in 1999 on the eve of the new millennium.

In 1998, accepted a higher-paying editor position with Canada's largest-circulation daily newspaper, The Toronto Star, while he continued to made progress in the interviewing, writing, organization and structuring of his Success Stories sequel project.

However, while the author's innovative approach was clearly working, conflicting visions and personality conflicts began to plague the new project and the partnership was dissolved in the summer of 1998, putting the *Success Stories* sequel project in limbo and its future in doubt.

Asked what prompted the falling-out, Davie humorously replies: "Ask yourself: Would you want to work with a driven egomaniac who always wants things done his way? No? You wouldn't want that? Well, neither did *he*."

The Birth of Manor House Publishing

With the partnership ended, Davie worked quickly to pursue his much larger publishing vision, by founding his own publishing house that same year, a home-based business he would run from his Ancaster residence.

Within months of the partnership break-up, Davie had searched for corporate names and was able to get his first choice approved for use. In the fall of 1998, he legally incorporated his new company as Manor House Publishing Inc. and established himself and wife Philippa as shareholders. Manor House was born.

First item on the agenda was to keep the Success Stories sequel project alive. After each and every one of the businesses he'd interviewed insisted they want to stick with him, Davie took over the project and worked fast to complete the book and meet the promised fall 1999 release date.

"I'd made promises, and I keep my promises," Davie explained. "The businesses were depending on me, other people were counting on me, and I wasn't about to let anyone down. My approach has always been to do what you say you'll do; to get the job done. Period – no excuses. Failure is not an option."

Enterprise 2000 and *Winning Ways*

Davie decided to give his *Success Stories* sequel a new compelling name and came up with *Enterprise 2000*, capturing the spirit of the new millennium. His highly original concept featured a New Year's baby on the cover, seated at a computer, signifying the importance of computer technology from infancy. That image was repeated endlessly on the computer screen against a boundless blue sky, denoting endless horizons of opportunity. Chapters included extensive information on how the educational system is preparing youth for the challenges that lay ahead. Business chapters similarly took a forward-looking approach. The book was a huge hit with the business community and general readers alike.

"We did so well with Enterprise 2000, that it bankrolled our general publishing program throughout our start-up years," recalls Davie, whose publishing house – assisted by Philippa and the couple's eldest son Donovan, aided by Sarah and Ryan – also publishes other non-fiction works, and fiction, including novels.

The *Winning Ways* series of business books followed, employing another original concept pioneered by Davie: changing the cover of the book to place each featured business leader's image on the front of all of their own advance-purchased copies.

Innovative approaches have continued with *The StreetSmart Marketer* and *Leaders* and Davie continues to break new ground with highly innovative and successful business book concepts.

Strategic Alliances

In 2005-2006, Davie took Manor House Publishing to a much higher level, forming strategic alliances with the giant distributor BookWorld in the United States and the United Kingdom-based distributor Gazelle Books. And when these strategic alliances are considered, the small home-based publishing house suddenly takes on a very, very big presence on the international scene.

These alliances allow Manor House to function as a virtual corporation, utilizing the large dedicated sales and marketing forces of the giant international distributors to effectively penetrate markets around the world.

While Manor House directly serves the Canadian market with coast-to-coast national distribution of its books, BookWorld services all 50 American states plus Central and South America, and through its affiliates, Australia and New Zealand.

Gazelle Books distributes through the UK, Europe, Russia, Asia, India, South Africa and many other nations. Alliances with the Amazon companies provide further worldwide market penetration via online sales.

"Our strategic alliances have made a huge difference in our ability to provide global marketing and distribution," Davie explains. "In essence, the distributors' sales forces are our own. BookWorld for example exclusively represents Manor House in its sales calls to Barnes & Nobel, Borders, Baker & Taylor, Ingram and various other booksellers. Gazelle also has a vast reach. Between the three of us, we've got our titles in just about every viable market in the world. Our books are in 100 different data bases of retailers; on numerous websites – virtually everywhere."

Davie is orchestrating further strategic alliances to build a global Manor House network to maximize worldwide success and catapult other business books into international prominence.

"Rather than start from scratch, I'm forging alliances to tap into the expertise, know-how and resources of complementary firms that integrate with our networked system and push revenue higher. What would otherwise take 10 years is instead happening now. We have a bright future ahead of even greater growth."

Greater than the current 50-100 per cent revenue growth rate?

"Well," Davie replies with a chuckle, "we can certainly try."

As Manor House Publishing continues to redefine business book publishing, entrepreneurs and business leaders of every kind are well advised to explore the possibilities offered by this highly innovative publisher.

"We're very open to considering, discussing and shaping project proposals," Davie asserts. "While we can't publish everything, our attitude from the outset is that we want to help as much as we possibly can. We're focused on finding ways to make a project work and bring it to a higher level; to make promising proposals a published reality."

Leadership Lessons:

1. Don't make assumptions: Take a proactive approach with clients to determine their needs and define your services.

2. Rather than start from scratch, forge alliances to tap into the expertise, know-how and resources of complementary firms and benefit from their years of experience.

3. Don't be afraid to make changes. There's always room for improvement. Small changes can make a big difference.

4. Let publishers deal with printers, bookstores, distribution. Self publishing is often a recipe for headaches with hidden or unforeseen costs. Unless all efforts fail, use a publisher.

5. Consider writing or being the subject of a book to establish yourself as an expert. Differentiate yourself from rivals.

6. Think outside the box and develop concepts that work better for you and your customer. Use your imagination.

7. Instead of forming a partnership, which can restrict your independence, hire freelancers and form alliances to get services you need while you alone control your business.

8. Once you have mastered one area of your business, expand into others to provide you with multiple income streams.

9. Train yourself to see potential where others do not. Make any modifications needed to help a flawed project succeed.

10. Utilize consultants to gain expert understanding/advice on key issues, rather than spend an excessive amount of your time trying to attain the same answers on your own.

At a Glance:
Mariposa Cruises:

Jim Nicholson

Title: President and CEO.

Claim to fame: Took over cruise line business, revamped it, re-branded it and turned into Toronto's premier cruise line. He is contemplating adding additional boats to the current fleet of seven. When all seven vessels are booked for an event, the potential passenger list can swell to nearly 2,000 people being entertained on a total of more than 34,000 square feet of space. Mariposa Cruises is entrusted with more corporate and private events than any other hospitality cruise operator in all of Canada.

Leadership: Hires superb personnel and lets them get on with providing expert services in a courteous manner. Staff hired on an as-needed basis to operate seasonal business. Sets high standards in customer satisfaction and strives to exceed them. Invests his own money and takes hands-on approach to running company with further financial assistance from silent investors.

Financial Data: Privately held multi-million-dollar company. Nicholson is one-third owner. Two silent investors each own a third. Thriving business, though seasonal, due to targeted marketing, partnering with other attractions, and experienced staff dedicated to creative a positive, memorable experience for clients.

For More information:
Mariposa Cruise Line's Office and Ships are located in the heart of the Toronto Harbourfront at the Queen's Quay Terminal Building (Intersection of York Street and Queen's Quay West)
Mariposa Cruises, 207 Queen's Quay West, Box 101, Suite 415, Queen's Quay Terminal, Toronto, Ontario, Canada M5J 1A7
North America Toll Free: 1-866-MAR-POSA (627-7672).
Local: 416-203-0178 Fax: 416-203-6627
Email: jimn@mariposacruises.com
Web: www.mariposacruises.com

5

Mariposa Cruises
Big Smoke on the Water

"Just imagine cruising Toronto's waterfront while entertaining your clients, motivating your employees, celebrating a corporate milestone... Mariposa is entrusted with more corporate and private events than any other hospitality cruise operator in all of Canada."
- Jim Nicholson, President and CEO, Mariposa Cruises

The lights of Toronto's booming skyline suddenly come into view as a clutch of international business executives gather on the cruise ship deck, cameras at the ready.

Clicking shutters punctuate the evening air, with businessmen clamouring for a memorable shot of one of the world's most gorgeous cities lit up in a vibrant glow.

This is the dramatic introduction to Toronto that will be forever etched in their memories. This is the big, bustling commercial centre they'll decide to invest in. This is the successful, dynamic image of Toronto they'll have in mind when making business decisions ranging from choosing suppliers to selecting meeting sites and expansion locations.

And *this* is the impressive view of Toronto they get *every* time they board a Mariposa Cruises vessel.

Toronto, affectionately known as The Big Smoke, is the prime attraction of such corporate cruises. It's the beckoning boom town rich with commerce and cultural charms, a centre of opportunity. And there is no more positive and memorable way to meet Canada's biggest city than on the deck of a Mariposa Cruises ship. You can choose from a variety of dinner cruises, private charters and Harbourfront tours. Just get aboard at their convenient location in the heart of the Toronto Harbourfront at the Queen's Quay Terminal Building on Queen's Key at the foot of Bay Street.

"It's a great experience," smiles Jim Nicholson, President and CEO of Toronto-based Mariposa Cruises. "Just imagine cruising Toronto's waterfront while entertaining your clients, motivating your employees, celebrating a corporate milestone or enjoying a private celebration... we also tour Hamilton's protected inner harbour and will sometimes venture out onto the open lake when the weather's nice and the Lake Ontario waters are calm."

Premier cruise line

Nicholson proudly notes Mariposa Cruises is Toronto's premier cruise line. "With Mariposa the possibilities are endless – including: weddings, theme parties, bar-mitzvahs, anniversaries, christenings, award ceremonies and various special events, especially business-oriented events."

"In fact," he adds, "Mariposa is entrusted with more corporate and private events than any other hospitality cruise operator in all of Canada."

Cindi Vanden Heuvel, Vice-president, sales and marketing, nods in agreement and points out that business cruises account for the bulk activity undertaken by the company's seven boats.

"At least eighty-five per cent of what we host onboard is corporate in nature," Vanden Heuvel observes.

"From employee appreciation cruise events, conferences and conventions, to client meetings, sales seminars and other business events – we do it all," adds Vanden Heuvel, originally from the

Stratford area and a former student who had worked summers on Toronto harbour, serving on yachts as a stewardess, deckhand and cook. She joined Mariposa in 1990 and went fulltime in 1996.

Vanden Heuvel notes the business that is Mariposa Cruises began in the mid 1980s and in 1987 took the name Mariposa Cruise Line, a nod to the flagship in the seven-boat fleet, the Mariposa Belle (formerly the Niagara Belle), which was built in 1970. The refurbished Mariposa Belle is still in use and is today the oldest boat in Toronto Harbour. To revamp its brand, the company recently shortened its name to Mariposa Cruises.

Nicholson, who also original hails from the Stratford area, invites corporations and individuals alike to: "Step aboard The Mariposa Belle and experience the magic of this legendary paddle wheeler."

"The charm of this ship creates an excitement to every event hosted on board – long established as part of Toronto harbour front history, the Mariposa Belle is the boat that started it all," he adds.

Nicholson, who runs the company with Vanden Heuvel, came in as president in 2003. In 2004 he bought into the company and became CEO as well as president, shareholder and owner-operator. There are also two other silent shareholder/investors.

"It's very much a seasonal business," Nicholson says of Mariposa Cruises. "June to September are our heaviest months and will climb to 145 people – a lot are temporary workers – during the summer months but drop to a couple dozen fulltime people in the off season when the boats receive additional maintenance. We have to earn enough to carry us through four months of the year when there is zero revenue coming in. But that's not a problem."

Diversified business career

For Nicholson, Mariposa Cruises is another chapter in a long and diversified business career that includes a prior 17-year career with Scotia Bank in the mergers and acquisitions field where he put together many millions of dollars worth of deals for a salary of about $45,000.

By age 37, Nicholson was a bank vice-president but felt restless and confident he could earn more in business on his own. He left the bank and began buying and selling companies, taking over some struggling printing companies to combine them with former competitors, creating a larger whole with a value exceeding the sum of its parts.

"I didn't know much about the printing business, but I did know how to read a balance sheet and I knew that operating costs had to be brought in line if there was to be any chance of profit," he recollects. Over the next two years, Nicholson reduced/right-sized the printing companies' workforces and further cut costs by consolidating all operations under one roof.

Beyond lending his enterprises some of his personal money that he'd borrowed to leverage an acquisition, most of the financing was accomplished not by borrowing money from banks but by getting the investors to take back mortgages and assume equity positions. By doing this, Nicholson was able to weather the 1990s recession and emerge with a much stronger business.

Assess opportunities

"You should always look at opportunities and determine if they're doable – and then see if you can reduce the risk in some way," Nicholson advises. "In this situation, there was no bank debt; it was all vendor take-back and the risk was minimized because the debt was on the shoulders of the former owners, all men in their 70s, who knew a leveraged buyout was their only chance of getting some value back out of their company."

"There was risk – but it was calculated risk," recalls Nicholson, who was then still in his late 30s with three young children under 10 years of age.

Nicholson held onto – and continues to hold – a one-third stake in the printing business with his silent investors. He also has a 10 per cent financial interest in Daytech Ltd., a company that makes bus shelters and he also owns some commercial real estate.

Just a few years into the new millennium, he and his silent investors decided to try something radically different – buying out a seasonal business in the hospitality industry – and so they acquired Mariposa Cruise Line Ltd., now re-branded as Mariposa Cruises.

"I really liked diversifying into the hospitality business with Mariposa," Nicholson recalls, noting there is "good name recognition now – and a solid balance sheet."

Vanden Heuvel notes the improved name recognition is partly the result of ongoing extensive advertising and marketing campaigns that promote Mariposa Cruises services worldwide.

"We market heavily on the international scene," she points out. "We're at many of the major trade shows and in tourism magazines and various publications throughout North America, Europe and Asia. We'll often offer tour packages with other attractions and accommodations."

Vanden Heuvel says Mariposa Cruises' extensive marketing campaign also includes direct mailings to corporate event planners, frequent contact with trade shows, associations and large corporations, Internet marketing, an updated website full of engaging features, and, customer surveys, efforts that often lead to repeat business and referrals.

"We've built a great data base – it's really all about knowing who your customer is and pleasing them. And we have phenomenal people who provide great service," she asserts.

Positive guest experience

"We focus on a positive guest experience and our team will take the time to make sure our guest feel comfortable – our team consistently provides courtesy and fun," she adds with a smile.

"We had been spending one per cent of sales revenue on marketing, business development and promotion. We're now spending up to five per cent – and we're looking at increasing this. Our marketing is paying very good returns on investment."

Nicholson nods agreement. "The experience people receive keeps them coming back and makes them want to refer us to their friends and business associates," he agrees.

"We provide the captain, the first mate, the crew and staff, the menu, the bar – your choice if its open or pay bar – and just about anything else you need or want," he adds. "We'll cater to your needs with chocolate fountains, cigar rolling, dinner and dances, massage services, theme parties, casino nights, live DJs, Dancing, hired entertainers, ice sculptures, decor services, balloon and flower arrangements... you name it."

Nicholson notes that three boats have their own Executive Chefs and service kitchens. These large kitchens cater to the smaller boats in the fleet. All vessels are licensed by the LLBO, Coast Guard Certified for appropriate passenger load capacities varying from 20 to 600 people; disc jockey, sound system and dance floors; washrooms; observation decks, comfortable seating; and professional coordinators to assist in the planning of your event. Rates depend on the date, menu and bar arrangements.

If organizing or booking a group, Mariposa Cruises offers a wide range of discounts, special rates and incentive packages for tour operators, schools, associations, businesses, travel planners, and other event planners.

For group discounts, Mariposa Cruises typically requires a minimum booking, signed agreement and deposit. Contacting a sales associate (416-203-0178) is the first step in customizing a special group rate and package (groups as small as 20 people can obtain group rates). Menu options and private narration services are also available.

Seven ships a-sailing

Mariposa Belle: Nicholson says the Mariposa Belle provides a good indication of the cruise experience available: "Renovated to meet the demands of social and corporate functions, she boasts three spacious promenade and dining decks," he notes. "You can dine indoors on the scrumptious meal prepared by our chef on board, then dance and take in the refreshing lake breezes atop the outdoor upper deck. From one of her characteristic promenade decks, your guests can appreciate the enchanting view of the city of Toronto. This charming vessel is ideal for mid to large size groups of up to 250 people. I'd invite anyone and everyone to let the Mariposa Belle create a little magic at your next event."

Captain Matthew Flinders: Or, if you have a larger group, why not consider the majestic 144-foot-long Captain Matthew Flinders, which has earned a reputation as a first class venue after hosting some of the country's high profile events. Named after the famous explorer who circumnavigated Australia, this vessel has charted new ground since arriving in Toronto's hospitality scene. The Flinders was custom built and lavishly outfitted with lush carpeting, warm woods and Danish brass. The intimacy of her two interior decks is complemented with spacious outdoor decks – and enough room on board for 600 passengers.

Northern Spirit: Another popular choice for comfortably entertaining large groups is the Northern Spirit. This 140-foot ship boasts three spacious decks from which up to 580 guests can enjoy the scenic cruise through Toronto's beautiful harbour. The two interior decks, with their abundance of natural light and subtle tones, create a spacious and luminous atmosphere. The third open deck provides you with the largest dance floor in the harbour.

Oriole: Or you can treat your guests to the beauty of the elegant 76-foot Oriole. The Great Lakes Steamship replica is decorated with varnished woods, wrought iron, polished brass and plush upholstery. The comfortable dining space is warmly lit by traditional brass coach lamps and features large picture windows and an oak stool bar. The two wrought iron staircases lead to the canopied upper deck where up to 185 guests can enjoy your choice of music. Often utilized for Mariposa Cruises' Sightseeing Program, the Oriole also hosts many mid-sized corporate and private functions.

The Torontonian: Like another choice? Why not experience the casual comfort of The Torontonian? Simply styled to suit a wide range of social gatherings, she is the traditional choice of Torontonians when it comes to having a great time on the water. The upper deck is fully enclosable but is usually open to refreshing lake breezes. The lower deck with its charming bar and buffet, provides up to 135 guests with a view of the city and its islands. The Torontonian is an ideal choice for touring the Island Parklands, intimate surroundings, smaller group functions and private parties.

Showboat: Another option is to relax and enjoy your cruise through the Harbour and Island Waterways aboard the Showboat. The Showboat's 62 feet boast charm and versatility. Along with an all-season upper deck, The Showboat's main deck is tastefully decorated with plush carpeting and comfy upholstered chairs. Bright picture windows ensure up to 90 passengers enjoy a scenic cruise through Toronto's harbour. With all the magic and magnetism of a paddle wheeler from the Old South, the possibilities for your event aboard are many. The Showboat also hosts many smaller to mid-sized corporate meetings.

Rosemary: Or, treat up to 70 guests to the luxury of the Rosemary, a gracious 68-foot yacht. When it comes to adding elegance and the perfect ambience to an intimate gathering or special event, The Rosemary is the classic choice. This yacht has played host to some of the most exclusive business and social functions. Tastefully decorated, the bright and spacious salon deck is an ideal spot for relaxed conversation. The fresh air decks provide the best views of Toronto's skyline. Whether it's an important corporate affair or a soiree put together on a whim, the Rosemary adds a touch of class.

Nicholson notes if all seven vessels are booked for an event, the potential passenger list can swell to nearly 2,000 people being entertained on a total of more than 34,000 square feet of space.

Three key elements

"Mariposa Cruises is clearly established as Toronto's leader in the hospitality cruise industry," he asserts, "because we provide three key elements required to guarantee the success of any event – the highest possible level of service on the best possible cruise ships in the best possible location. Our reputation has been built on our excellent food, our clean, well-maintained ships and our courteous professional staff."

Nicholson provides still more reasons for choosing Mariposa Cruises: "Having commenced operations in 1987, we are now one of the largest cruise lines in North America – but we are constantly striving to improve our products and services to meet our clients' needs. We've had the pleasure of hosting thousands of events and greeting millions of guests. Our event coordinators will organize every detail for you, and our uniformed hospitality and navigation crew will cater to the comfort of your guests on board. Great meals are freshly prepared on board by our chefs. We also have the flexibility to customize and provide any menu or buffet-style meal for your event."

"You'll also find professional sound systems and disc jockeys," he adds, "and we're fully licensed by the LLBO and

Coast Guard Certified. Best of all, our ships are located in the heart of downtown Toronto with easy highway access, plenty of parking, 5 nearby major hotels, TTC service and only one block from Union Station, CN Tower, Air Canada Centre, Rogers Centre and many other attractions."

In addition to custom corporate events and special occasion group bookings, there are also regularly scheduled cruises available for individuals or groups of various sizes.

There are available scheduled cruises including: a 1-hour Narrated Sightseeing Tour, a 2-hour Casual Luncheon or Sunday Brunch Cruise, or a 3-hour buffet-style Magical Dinner Cruise with disc jockey and dancing on 50 available dates a year (1-866-627-7672 or 416-203-0178 to book cruises). The narrated cruises offer cash bars, indoor and outdoor seating, washrooms, and views of the Toronto skyline the Toronto Islands, Island Airport, Harbourfront, CN Tower, SkyDome (Rogers Centre), Air Canada Centre and the Gibraltar Point Lighthouse. Your boarding point is conveniently located on the harbourfront in the heart of downtown Toronto, with ample parking, TTC Harbourfront LRT service. It's within two Blocks of Union Station, and five major hotels.

"We offer five regularly scheduled harbour tours a day," Nicholson notes, "where you can cruise through Toronto's scenic Harbourfront and Island parklands for a very enjoyable time. Our charming ships are first class venues, tastefully decorated, air-conditioned, fully licensed, and versatile and spacious enough to accommodate groups and functions from 20 to 600 patrons. Ships may be chartered or individual decks hired for smaller parties and functions."

Film and TV industries magnet

Vanden Heuvel says it's not surprising that film and television industries are particularly fond of Mariposa Cruises.

"Conveniently located in the heart of downtown Toronto, Mariposa Cruises welcomes the Film and Television Industry and Photographers to our vessels for shoot locations," she notes. "The seascape ambience of the Harbourfront, surrounding Yacht Clubs, City Centre Island Airport, Great Lakes Waterfront, the beautiful surroundings of forested Island Parklands and our various charming ships afford a variety of settings on their own."

"Coupled with our location amidst the stunning modern cosmopolitan, yet mixed with traditional old-world architecture of the Toronto Skyline, Mariposa Cruises offers a great range of versatility as a filming and photo shoot venue – and we're very popular for shoots and wraps," Vanden Heuvel adds, noting the producing of movies, videos, media development, advertisements and television programs is a thriving business in Toronto, now known as the third largest exporter of film and TV production in North America after New York City and of course Los Angeles. Also known as "Hollywood North" Toronto has hosted hundreds of Hollywood features, commercials, animation productions and television series for many international markets.

Mariposa Cruises has entertained several productions including films, TV episodes, commercials, news broadcasts and advertisements as a filming locale. Past features include: Ontario Lottery and Gaming Corporation (Lotto 649 commercials); and TV/film productions Wild Card, Kung Fu (Warner Brothers); Anne of Green Gables (Sullivan Entertainment); and the popular TV series Due South (CTV).

Leadership programs

In addition to serving the film and television industries and to providing all corporations with an array of cruise options and great venues for conferences, meetings and seminars, Mariposa Cruises also offers an unbeatable setting for business leadership programs – including a L(earn)2 program hosted onboard, which is based on saving the Titanic. And in devising ways the Titanic might have been saved, the participants simultaneously come up with ways to save their conference or company.

Testimonials

Mariposa Cruises clearly has some unique and enjoyable meeting or conference venues for mid-size corporate events, conferences, product launches, sales meeting, seminars or workshops.

Not surprisingly, the various services offered by Mariposa Cruises have drawn rave reviews. Testimonials include:

Ontario Egg Producers Marketing Board:
"The cruise flowed smoothly, the meals were delicious and the buffet tables were beautifully presented.

It has, again, been a pleasure working with Mariposa Cruise Line.

Your professionalism and special attention to details are truly appreciated."

Dial One:
"The night of our Christmas cruise probably the best night of the year to be on the lake.

The weather was perfect, the scenery was breathtaking and the skyline of Toronto looked like a postcard.

The staff gets a five-star rating.

Dial One has another definition for the word perfect – "a Christmas Cruise aboard the Captain Matthew Flinders."

Tower Perrin:
"I have been on many dinner cruise events over the years, from a ship on the Nile to several on the California Coast.

This experience was by far the best. The ship was outstanding and just right for our size; the crew was extremely friendly, helpful and proportionate to our group; the food was excellent, well above the standard shipboard fare; and the entertainers engaged made for a delightful evening."

TD Canada Trust:
"For all the years I have organized company functions, I have never had such pleasure in dealing with a company and its employees as I did in my dealings with Mariposa Cruises.

We would like to extend our sincerest thanks for having such wonderful staff members who went out of their way to make our day the success that it was, and for such a lovely, well kept vessel."

Travel Adventures: "Your service is a perfect match for our student tour groups.

Your staff has always demonstrated the highest levels of professionalism and expertise.

Our experiences while visiting your facility have been very pleasant.

We are always greeted by friendly, competent employees."

That's just a small sampling of the many words of praise splashed on Mariposa Cruises.

Nicholson knows he has something truly special in this remarkable business venture/cultural experience.

He invites business executives everywhere to consider a floating venue that showcases the very best Toronto has to offer: "For your next corporate function, business meeting, employee get-together, product launch, company event, family outing, event or wedding, charter one of our vessels to ensure your event is a unique and memorable experience."

Leadership Lessons:

1. Always look at opportunities, assess them, and determine if they're doable – and then see if you can reduce the risk in some way while still deriving benefit.

2. Taking risks is often necessary to achieve success, but make sure that each risk you take on is a calculated risk in which the odds of success are in your favour.

3. Leadership involves being the best in several key areas. For example, Mariposa Cruises offers the highest possible level of service on the best possible cruise ships in the best possible location.

4. Apply targeted marketing to wisely aim your advertising dollars directly at those most likely to use your services.

5. Leadership is also about translating ideas into action, communicating context, taking leadership, and creating innovative solutions as a team.

6. Providing great customer service is the surest way to build your business through referrals and become a leader in your field.

7. If undertaking a leveraged buyout of another company, strive for vendor take-backs of mortgages or debt to minimize costs and risks.

8. Increase your marketing to grow your business.

9. Partner with other companies that complement your own and derive mutual benefit from the synergies that result.

10. Don't cling to an outdated or ineffective message; re-brand your company to convey a clear message in a memorable manner.

At a Glance:
The StreetSmart Marketer:

Michael Hepworth

Title: President of Results Exchange Group and founder/president of all StreetSmart Marketer enterprises.

Claim to fame: Established the personal brand StreetSmart Marketer, along with proven marketing strategies that have helped thousands of clients achieve success. Michael Hepworth developed a seven step process that starts with understanding who you're targeting and articulates what you do to resolve the potential clients' problems, discomfort or business pain; it involves working with people to overcome common mistakes they unwittingly make – to help them move ahead at a rapid pace.

Financial Data: Privately held. Michael Hepworth and his wife Madeline are equal owners of his four-year-old rapidly growing business. Michael Hepworth's previous experience includes managing and operating several prior successful companies, which he subsequently sold at a profit, allowing him to live a financially independent lifestyle, doing whatever he wants to do, whenever he wants to do it.

New Book: *The StreetSmart Marketer* (Manor House Publishing, Ancaster, Ontario, 2006: 144 pages, $24.95 CDN/$19.95 US)

For More information:
Contact Michael Hepworth:
Email: streetsmart@streetsmartmarketer.com
Phone: 416-204-0353
Website: www.streetsmartmarketer.com

6

The StreetSmart Marketer
Down to earth advice that really works

"StreetSmart Marketers engage in *You Marketing*. Their total focus is on selflessly serving their customer, helping them overcome problems, educating them how to be better buyers and raising their perception about how they can help. They relate to their customers, not so much because they have the best products and services, but because they understand them first as people."
- *Michael Hepworth, Founder/President, Results Exchange Group and founder/president of all StreetSmart Marketer enterprises.*

Does your sales pitch fail to focus on a potential client's needs? Do you try to close a sale before your prospects understand why they should be interested? Is your marketing more about you than it is about your customer?

If you answered yes to any of the above questions, you stand to benefit from the wisdom of marketing guru Michael Hepworth, known to clients, fans and followers as the StreetSmart Marketer, a trusted source of low-cost strategies to help business owners work less but make more profit.

"During the course of a week I usually meet several people who hope to sell me something – what astonishes me is how few of these people ask questions to find out what I am interested in or worried about," notes Hepworth, who is also president of the Toronto-based Results Exchange Group, and author of the new bestseller *The StreetSmart Marketer: Book One: 11 Keys That Unlock the Secrets to Rapid Growth In Your Business* (Manor House Publishing, Ancaster, 2006, $24.95 CDN/ $19.95 US).

Hepworth, throughout decades as a highly successful business leader, has encountered numerous ineffective sales efforts.

"Most start with a pitch aimed at impressing me with their success. I hear about all the amazing customers they have... Then they tell me all about the wonderful technology they have, or how marvelous their products or services are... by this time I have generally lost interest. It is such a shame that these otherwise competent people shoot themselves in the foot and most lose the opportunity to sell me something because they are so focused on themselves and have little understanding or feeling for what I might be interested in. I call this *Me Marketing*. It is a lot more common than most of us realize and destroys more sales than almost any other mistake," observes Hepworth who has developed low-cost strategies to transform marketing failures into successes.

Hepworth, 57, likens such *Me Marketing* failures to the boastful approach taken by an arrogant suitor: "It is like bragging about your possessions or achievements in the hopes that you will impress your date. In most cases all it achieves is boredom, or even worse, it turns your date off completely. This approach has the same effect with prospects."

You Marketing

He asserts: "StreetSmart Marketers engage in *You Marketing*. Their total focus is on selflessly serving their customer, helping them overcome their problems, educating them how to be better buyers and raising their perception about how they can help. They relate to their customers, not so much because they have the best products and services, but because they understand them first as people."

The *Me Marketing* versus *You Marketing* approach is just one of the many subjects covered on Hepworth's online *StreetSmart Marketer* newsletter, filled with tips and advice and available free of charge on his website: www.streetsmartmarketer.com. And, as Hepworth notes, success is built on positive relationships: "The truth is, people buy from people they like, and we generally like people who are interested in us and our needs. In most cases we are completely selfish and will only engage, if we find something or someone to be interesting. Successful salespeople quickly learn that you can't be interesting unless you are interested."

Hepworth adds that timing is everything when it comes to introducing the product or service to be offered. "StreetSmart Marketers know that the best way to demonstrate interest is to ask insightful questions. They provide insightful answers in their marketing communications, showing customers how they are the logical choice to solve some burning problem. Once they understand these challenges they use their knowledge and skills to raise perception about potential solutions. Only then do they introduce the product or service and what it can do for the client."

"StreetSmart Marketers practice *You Marketing* by making sure that all their marketing materials focus first on customer issues before their capabilities and competencies," he adds. "They never pitch without first building rapport. Above all they know that customers don't buy from them because they have the best products and services, but because they understand them as people. Their marketing is designed to demonstrate that understanding."

Sought-after Speaker

Not surprisingly, Hepworth's proven expertise in crafting effective sales pitches and in other marketing matters has made him a sought-after speaker, workshop host and seminar presenter. With clearly presented self-help messages, Hepworth's engaging and informative new ***StreetSmart Marketer*** book has further enhanced his stature as one of the nation's leading marketing experts. And a seemingly endless list of testimonials also speaks to his ability to help businesses succeed on a global basis.

How Michael Hepworth became such a highly regarded marketing expert is a story that begins in Africa. Although born in England, Hepworth moved with his parents, Jack and Heather Hepworth, to what was then Rhodesia (now Zimbabwe) before he was a year old.

Hepworth recalls an idyllic childhood of endless summery days amid natural splendour. "It was wonderful growing up there; a very free and easy lifestyle – like growing up in paradise – great weather, wonderful times..."

At age 18, Hepworth was conscripted for mandatory service in the Rhodesian army, compulsory for all males of that age. The normal time commitment was an initial nine months of basic training followed by six weeks of service annually. But Hepworth soon found the army demanding a much larger share of his time.

"There was a war going on – the Rhodesian armed forces were fighting local guerrillas – and as the war continued, we were spending six months a year in the army."

Around that time – roughly 37 years ago – Hepworth also began his marketing journey when he studied marketing, between army stints, at Salisbury Polytechnic in what is now Harare in Zimbabwe. "I had heard that marketing was truly the business of business and if you wanted to be successful in business, marketing was a great place to start," he recollects.

"I naively thought that what I learned in school would set me up for a life of success," he adds. "Imagine my disappointment when I learned in the workplace that although I had a good academic grounding, much of what I'd learned was of limited practical use. This was especially true in small businesses with limited marketing budgets. With the impatience of youth, I quickly became disillusioned with starting at the bottom, I gradually moved away from marketing into sales, where if you were good there was virtually no limit to what you could earn and achieve. Everything I learned had some practical application. I readily found ways to apply what I was learning."

Strategic Business Services

And all of these achievements came while Hepworth was spending literally half his time in the Rhodesian army. After serving in the army for eight years, Hepworth realized he wanted out. "I began to question what I was doing; what we were fighting for; and I began to see it as a futile war – so I left when I was 25."

Hepworth's education led to workplace experience and expertise. From 1986 to 1988, he worked for the Wilson Learning

(South Africa) company, much of this time as Managing Director of Strategic Business Services Ltd., the consulting division of WLSA.

He had started SBS with a mandate of helping WLSA sell more learning programs. "It was designed as a change management company to help firms go through change and execute change," Hepworth explains. "Often change gets initiated but a company doesn't follow through – one of the tools in executing change is training and we provided that through our programs. SBS was specifically started to grow sales through the addition of value added services."

This highly productive two-year period was marked by loss at its outset in 1986 with the death of Hepworth's father, a retired civil servant and inspiration to his hardworking son.

Despite this loss, Hepworth managed an impressive string of accomplishments. While with SBS – "the first company I ever ran" – Hepworth built a new viable business unit based on a new computer based facilitation tool for use on strategic planning projects; conducted strategic planning and organizational change projects for leading South African organizations; developed marketing plans with the tool that enabled WLSA consultants to access highest levels of executives in target companies; and, took sales from zero to 60 per cent of corporate revenues in less than two years.

Canada – and Forum Corporation

The war would continue a few more years, but without Hepworth. "I came to Canada in 1988 – I'd been to Canada a few times before and I liked the values of the country and the great outdoors. And I had a sister here – that made it appealing as well."

His sister Gill still lives in the Toronto area, just a 10 minute drive from the Willowdale bungalow Hepworth and wife Madeline have called home for the entire 18 years he's lived in Canada. The couple are equal owners of his four-year-old, home-based, multi-million-dollar business. The couple have two sons, Simon, 31, and Adam, 29, newly wed following a wedding in November 2006 in Capetown, South Africa. And there are grandchildren Mia, 5, and

Hanna, 9 months. Hepworth's mother, 96, still lives in Zimbabwe "so I go back to visit her whenever I can."

On arriving in Canada and Toronto in his mid-30s, Hepworth joined the Toronto area offices of the Boston-based firm Forum Corporation Inc. He stayed from November 1988 through to January 1990 with the large consulting and training firm serving corporate clients. Hepworth, as Director of the Canadian Consulting Practice and member of the board, was responsible for Canadian sales of Forum's consulting products. He also handled consulting projects aimed at helping large Canadian firms become more customer focused.

Hepworth + Company

But Hepworth soon began feeling restless. He wanted to be his own boss. Fifteen months after arriving at Forum he left and founded Hepworth + Company Ltd., a leader in call centre, customer loyalty and customer retention consulting.

Hepworth started the business from scratch in February 1990, funding its start-up and growth from cash flow.

He served as President and CEO from start-up to July 2001, establishing the firm to be a high end consulting boutique and a recognized leader in the fields of customer retention and customer loyalty.

"It was the first time I'd been completely on my own," recalls Hepworth, who can appreciate what entrepreneurs go through, "because I've experienced it myself five times – I'm a serial entrepreneur."

But for a young man in a new country leaving the security of a conventional employer is fraught with added risks.

"It was a scary time," Hepworth recollects. "I was in foreign country with no established financial support mechanisms to fall back on, no networking systems to turn to, and no margin for error; there could be no possibility of failure – failure was not an option. It's really pretty scary when you don't have the support infrastructure that someone here with a family has – instead, you have to rely on own resources to make things happen. Fortunately, from my years in boarding school and in the army, I had learned to be resourceful, so I did everything possible to make my new venture work, and I was very determined to succeed."

Hepworth soon found that the old marketing approaches he'd been taught made little sense in the context of his new venture.

"The stuff they teach you in school doesn't have much relevance in the real world of a small business entrepreneur – much of what you're taught is based on the assumption you'll be working for a large corporation with a big marketing budget," he observes.

"Although I had a marketing background, I certainly didn't have a big marketing budget. I had to find low-cost, low-risk approaches as I didn't have a huge company. I kept the costs as low as I could and went through a constant trial and error process to make it work – and this remains a continuous process to this day – I'm always finding better low-cost ways of doing things for improved results."

Hepworth steered clear of costly advertising and other expensive marketing vehicles and instead marketed the business inexpensively through speaking engagements, writing articles and holding customer forums.

He also developed a proprietary customer satisfaction measurement system that enabled customers to calculate the bottom line impact of poor customer service.

Low-cost Measures

All of these low-cost measures raised the profile of his company and attracted a burgeoning list of prominent clients.

Indeed, he built a large blue chip North American client list to include names such as: Royal Bank, Scotia Bank, Kellogg Brown & Root, Halliburton, Pratt & Whitney, AT&T. Also did business in Netherlands and the United Kingdom.

Hepworth finds he's learned at least as much – possibly more – from his failures as he has from his successes.

"I have read hundreds of marketing books and, attended more marketing courses than I care to remember," Hepworth notes.

"I've had almost as many failures as the hot breakfasts I've enjoyed. But each failure has taught me some small lesson which I have added to my experiential tool box. I have also had many successes, among them helping several businesses double in size in a matter of months."

The necessity of having to find effective marketing techniques that cost little to implement proved to be the catalyst for a life-long career in developing low-cost marketing strategies.

"One of the most common misconceptions is that marketing has to be very expensive – I've found that's not necessarily true and that sometimes the least expensive approaches can yield the best results."

And, speaking of great results, the last three years Hepworth served as CEO of Hepworth + Company Ltd., he produced a greater than 25 per cent growth rate.

He then managed the sale of the business and merger of the firm into Boston-based Seurat Company for an undisclosed sum. The acquisition allowed Seurat to expand into Canada.

Hepworth became a managing director at Seurat and gained access to the expert resources of Seurat's consulting services while Seurat gained access to an established client base outside of the U.S. and took on an active sponsorship role in the Customer Contact strategy Forum (CCSF), which was co-founded by Hepworth.

CCSF has worked with more than 80 blue chip companies to develop call centre strategies.

Customer Contact Strategy Forum

Hepworth served as President and Founder of Customer Contact strategy Forum (CCSF) from June 1999 to April 2002 and established CCSF as a leader in its role as an exclusive North America information exchange for call centre executives.

He was responsible for developing the concept; finding strategic customers willing to provide funding to get the business off the ground; and, for building it into a global business.

Hepworth started the firm as a marketing idea in conjunction with IBM, Nortel, PeopleSoft, Drake and Bell Canada.

Then he drove it from an effective marketing idea, into a viable business that continued to generate significant leads for the business sponsors. And he built the business into the leading forum for call centre executive discussion and problem-solving.

In short order, Hepworth grew the membership to more than 260 vice-presidents across North America and Australia. The members had combined buying power of more than $8.7 billion.

In 2002, Hepworth successfully also managed the sale of the business and merger of the firm into Seurat Company.

Over the past 20 years he built three successful businesses, generating more than $39 million in sales. He then sold each one, cashing out with enough funds to live in style the rest of his life.

Results Exchange/ StreetSmart Marketer

"But instead of just retiring — something that would have bored me into an early grave — I decided to begin sharing my business-building secrets with a select group of clients, who are now building fabulously successful businesses of their own," explains Hepworth, who, when he's not marketing or consulting, enjoys spending time with his family and engaging in such recreational pursuits as boating, physical fitness, cooking and travelling to foreign countries.

Within a few months of selling CCSF, Hepworth founded the Results Exchange in August 2002 and served as the company's President to this day. The Results Exchange specializes in creating revenue-generating strategies for entrepreneurial businesses and also does contingency based marketing consulting.

"Results Exchange is the legal name of company at the moment and StreetSmart Marketer is the brand," explains Hepworth, whose complementary newsletter reaches avid readers in 35 countries every two weeks, and whose current businesses have helped many firms succeed on a global basis.

"But it's likely we will change the name and the company will be known as The StreetSmart marketer as its adopted brand," Hepworth notes. "It really all began when I got together with a group of friends and colleagues and we came up with name brand StreetSmart Marketer over a coffee during a brainstorming session." And so The StreetSmart Marketer was born.

Hepworth realized that much of traditional marketing seemed aimed at big corporations with equally large marketing/advertising budgets. As well, the marketing advice and theory taught in schools seemed somehow removed from the reality and real life experiences of smaller businesses.

In reaching out to small business owners, Hepworth focused the advice on aiding entrepreneurs. And the advice itself was obtained from an array of creative and innovative marketing experts – including Hepworth himself.

Indeed, Hepworth has a good deal of experience in thinking outside the box to achieve success. In each business he developed, he identified underserved areas of the market. Then he developed effective strategies to take advantage of these opportunities and designed the organizations to take advantage of each opportunity. Most impressively, he produced steady year over year growth by changing and adapting marketing strategies as the market evolved.

He created consistent growth through high-yield marketing strategies and programs. Simply put, Hepworth practices what he preaches: He thrives in environments where creative problem solving is required – and finds non-linear solutions to difficult-to-solve problems.

Strong Sales, Marketing and Leadership Skills

Hepworth has also developed proven leadership skills over the years, leading to the achievement of corporate vision and strategy. He's managed the merger of acquired businesses with the new owner's existing business.

He has an extensive personal network of effective business contacts, and believes in setting realistic expectations, building commitment from the team to achieve the goals and then driving the team towards success with pragmatic solutions aligned with the strategy

His advice is extremely valuable, coming from someone who has been there, done that, and has the battle scars and experience to prove it. Remember, he built four profitable businesses, from the ground up. Each business was built from cash flow, without the benefit of invested capital. Each business became a market leader in its particular segment.

Then there's his experience in mergers and acquisitions: he sold two of the businesses to US based investors and he's advised other entrepreneurs interested in selling their businesses on structuring more attractive deals.

Hepworth also brings 20 years of successful sales and marketing experience to the table.

He's not only built and managed four successful sales and marketing functions in professional services organizations, he's also sold services to large multinational organizations, such as banks, telephone companies and insurance companies.

Hepworth also built strong "C" level (chairman, CEO, CFO, COO) relationships. And he built a strong brand through effective marketing in each organization. Hepworth recruited and developed an effective sales team, implemented sales and marketing strategies in all four companies, and produced consistent year over year growth in sales, in each organization.

He's developed market leadership through a strong customer driven culture in each organization and operated as a consultant to Fortune 500 and Financial Post 500 businesses for over 20 years.

As a public speaker he's lectured at the Schulich School of Business on Advanced Customer Satisfaction Management and Entrepreneurial Marketing, along with various corporate conferences for Bell Canada, St Lawrence Cement, ScotiaBank and other major corporations.

StreetSmart Marketer book educates entrepreneurs

Hepworth's *StreetSmart Marketer* book, reasonably priced at $24.95 Canadian/ $19.95 US, also provides a low-cost means of obtaining a great deal of helpful, well presented information in a convenient single package that can be referred to again and again.

With its engaging storyline and expert advice, the book is also taking Hepworth's business to another level while educating a growing number of entrepreneurs on the joys and successes of low-cost, low-risk marketing approaches that are highly effective.

In addition to his *StreetSmart Marketer* book and newsletter, his writings have been featured in the pages of *Direct Marketing News* and *Canadian Business* magazine.

As Hepworth immersed himself in the entrepreneur's world, he saw that the small business owner/entrepreneur – despite being the leading creator of jobs and prosperity – was being overlooked.

A vast niche market was not being served – and he saw an opportunity and stepped into the vacuum.

"So much of the marketing advice that was out there really didn't relate the small business owner, the entrepreneur," Hepworth notes.

"What was needed was an approach based less on academic theory and more on down-to-earth, low-cost, practical approaches – something less book-smart and more street-smart."

And so The StreetSmart Marketer was further defined.

Low-cost, low-risk marketing

"StreetSmart Marketing is low-cost, low-risk marketing for owner operated businesses – it's mainly for smaller businesses," notes Hepworth, whose ***StreetSmart Marketer*** electronic newsletter, in keeping with the low-cost theme, is offered free of charge to subscribers who provide nothing more than an email address to start receiving helpful advice at no cost.

In the pages of ***The StreetSmart Marketer*** newsletter, the free subscribers can find many of Hepworth's simple but highly effective strategies, techniques, and secrets his clients have used to achieve remarkable results. "And when you start putting them to work in your own business, you'll immediately see substantial increases in both your response rates and your sales and profits," Hepworth asserts.

"And it's all based on what I've learned in the trenches over the last 20 years," he adds. "Every idea is battle-tested and proven in the real world in real businesses — from traditional brick and mortar businesses to direct response and internet businesses."

"More importantly," Hepworth emphasizes, "the techniques I share are easy to use and won't cost you an arm and a leg to implement. I show you how to improve the effectiveness of the advertising and marketing you're *already* doing, so you'll get three, four, or even five times as many leads or sales for the same effort."

Same cost – better results: An unbeatable combination. But there's more:

"Better yet," Hepworth smiles, "you'll begin to experience how much fun business can be when you don't have to work 24/7 and then some, just to ensure a steady supply of new customers. It really is amazing how enjoyable business can be when your advertising is working, your sales and profits are growing, and you can finally take some time off to spend with family, pursue your favourite hobby, or simply recharge your batteries.

Indeed, a growing number of free subscribers are finding the expert advice is helping them succeed on a global basis.

"Many clients have quite wide reach, around the world," Hepworth concurs, "and we have subscribers to the newsletter in so many countries, it really is global in scope. The newsletter is my primary communication vehicle to provide free information to clients and potential clients. Although we sometimes offer products and services for sale, the newsletter provides value and is not pitching product for the most part."

While most of his clients are small and are not familiar names to most Canadians – yet – there some corporate giants in the mix, including Hewlett Packard, which is offering his program to a growing number of resellers. RBC is offering his programs to their mortgage brokers.

The TD Bank and Royal bank have used Hepworth for speaking engagements for their entrepreneur clients, and many big businesses are discovering that the principles that have helped so many small businesses can also be applied to larger firms with great results.

For Hepworth, the acquisition of marketing expertise has been a journey of lifelong learning, of constantly making new discoveries and finding better ways of marketing.

Entrepreneurial Marketing

"Many of the discoveries I made were driven by my naiveté and unwillingness to accept the status quo," he recalls. "One of my most profound discoveries has been the difference between entrepreneurial marketing and traditional marketing. The two approaches are so different that you have to forget what you know about traditional marketing to be effective in entrepreneurial marketing."

"I quickly figured that although cold calling could be effective in reaching a prospect, it was an inefficient process for building a business," Hepworth recalls.

"It couldn't produce what I needed: a steady stream of new clients that would call me, eager to buy my services. Some people told me there was no alternative; you had to just get on the phone and make those calls. To some extent that's true. I still believe that the most powerful marketing weapon we have, is one-to-one selling. However, the key issue in almost every business is how to get the phone ringing so you can do some one-to-one selling."

Hepworth immediately began developing programs that were as entrepreneur-friendly as possible.

"I began to look for marketing that was inexpensive to implement and would not involve having to work too hard. Over the years this has developed into a philosophy of marketing, that I call now entrepreneurial marketing."

Hepworth notes there are numerous areas in which entrepreneurial marketing differs from traditional marketing.

Time, Imagination and Energy versus Money

"Traditional marketing requires a significant investment of money, whereas entrepreneurial marketing requires the investment of time, imagination, energy and knowledge – but not necessarily much money," Hepworth explains, noting this is one area of major difference between traditional and entrepreneurial marketing.

Hepworth elaborates: "Entrepreneurial marketers are patient above all else. Because money was scarce, I have used what I now call entrepreneurial marketing to build several businesses from scratch with no initial capital, just cash flow. In the beginning, sales tend to be slow until your marketing efforts build momentum. However, once up to speed, low-cost marketing activities such as referrals, joint ventures and public speaking, can keep new business flowing with very little effort and almost no cost."

"Entrepreneurial marketing is a better approach for businesses with limited funds; well executed, it can inexpensively produce significant results," he adds.

Another major area of difference between the two styles of marketing has to do with the size of business venture: Traditional marketing is geared to big business with deep pockets and plenty of wiggle room. But entrepreneurial marketing is more geared to small to mid sized business because it requires less investment and poses less risk. "This does not mean that large businesses should not or cannot do entrepreneurial marketing," Hepworth points out. "Quite the contrary: I believe that all businesses could save large sums of money by adopting some of the principles in entrepreneurial marketing."

Profit the Key Measure of Success

Hepworth observes that yet another difference between entrepreneurial and traditional marketing is that entrepreneurial marketers measure their success with profits "and every program has to pay for itself or you stop doing it."

In contrast, he notes that traditional marketing is more frequently measured by sales, response rates and leads.

"Large businesses are obviously interested in profits," Hepworth acknowledges, "but generally do not ascribe their profits to a specific marketing program. I believe that marketing needs to be accountable and that each program needs to be measured just as you would a sales person. If it's profitable, you keep doing it. If it is not profitable, you modify it until it is; or you stop and do something that is profitable."

Nor do entrepreneurial marketers have the luxury of time and resources to turn money-losing mistakes into success in the long term, Hepworth observes. "Traditional marketing is largely based on years of experience. Thus it takes years for anyone to become a successful marketer in this field. Big dollars and splashy campaigns often cover marketing errors that would severely hurt a small business that is unable to saturate the market."

"In contrast," Hepworth adds, "entrepreneurial marketing is based on an understanding of human behaviour. Entrepreneurial marketers know that purchase decisions are made in the unconscious mind and that you can best work on the unconscious mind by repetition. For this reason they communicate frequently with customers via any medium that makes sense. They draw them in slowly, building trust and rapport as they go, teaching prospects how to buy and providing value every step of the way."

Hepworth says successful entrepreneurial marketers focus on the core message. "Entrepreneurial marketers quickly learn that to grow they must maintain focus. They also learn that a growing ego can quickly result in a loss of focus. Entrepreneurs are notorious for going in multiple directions only to discover the negative consequences when business begins to decline.

Hepworth notes different growth patterns are also a factor separating the two forms of marketing: "Traditional marketing focuses on linear growth; through acquiring one customer at a time. Entrepreneurial marketers find ways to grow geometrically: They look for alliances that will create a constant stream of new business through referrals and endorsements. They also look for ways to increase the size of their sales by up-selling and cross-selling at every opportunity. They increase the size of their business by offering back-end products and services to satisfied customers."

Make the Easy Sale First

As Hepworth notes, traditional marketers rarely look to making the easy sale but instead count on their established brand to make things happen. The brand stands for something that the purchaser can trust. This powerful brand is relied on as a way to gain access to customers.

Entrepreneurial marketers usually lack an established brand and must work harder to make sales happen. While traditional marketers lean on their brand to bring customers to them, entrepreneurial marketers are focused on meeting one-on-one with prospects, Hepworth explains. "To do this, to meet one-to-one, the entrepreneurial marketer must find superior access vehicles to open doors for them. Usually this can be achieved by offering some kind of useful information that buyers need. It can be done in special reports, executive briefings or it can be simply provided over the phone. The primary goal is to educate the buyer about becoming a better buyer."

Hepworth observes that, not surprisingly, entrepreneurial marketers are fervent in their follow-up, knowing that 68 per cent of all business lost is as a result of apathy after the sale. They continuously follow up, never letting a prospect have enough time to forget about them.

"In contrast," Hepworth adds, "while traditional marketers *talk* about staying in touch with customers, they focus more on new business and invest money in seeking new clients more than they focus on retaining existing clients and nurturing prospects."

"And while traditional marketers generally simply use their financial resources to try to obliterate the competition, entrepreneurial marketers use other people's assets to reach their customers. They form alliances with businesses that have the same prospects as they do. In this way, they can take advantage of the huge investments already made in developing clients and infrastructure."

Wide Range of Marketing Tools

Interestingly, most traditional marketers use only a handful of marketing tools, mainly found in the equally traditional media, such as newspaper, magazine and radio/television advertisements. Yet as Hepworth notes, entrepreneurial marketers have more than 50 marketing tools – and most of them cost nothing to implement.

"Entrepreneurial marketing tools include: testimonials, joint ventures, strategic nurturing of prospects, referrals, back-end selling, cross-selling, up-selling, down-selling, customer education, public speaking, writing articles, pre-programming purchases, endorsements, personal communication and developing irresistible offers," Hepworth explains, adding that combinations of tools often work better than individual tools on their own, as each tool supports the other until its impact is felt. "You can combine direct mail, with advertising, public speaking, telemarketing and a host of other tactics, never relying on one alone to support growth plans."

"Another key different between traditional and entrepreneurial marketers," Hepworth adds, "is that traditional marketers develop messages aimed at groups they call markets, while entrepreneurial marketers develop messages aimed at individual prospects and customers. Simply put, entrepreneurial marketing is about dialogue with customers. Entrepreneurial marketers know that by talking to and listening to customers they will get their best ideas for improvements and for new products. In sharp contrast, traditional marketing is a monologue directed at customers."

Hepworth elaborates that at its core, entrepreneurial marketing is *you marketing*. "It talks to prospects about the problems they are facing, the issues that keep them awake at night – and it answers their unspoken questions."

"In contrast, traditional marketing is *me marketing*," Hepworth explains. "It's all about how great the company's products and services are, how effective its people are and how it has the biggest, best and most expensive equipment. The customer's needs don't factor into this."

Hepworth says that entrepreneurial marketers are givers. "They know by giving free services, information, samples and by educating their prospects, customers will learn to trust them and many will buy from them. They reverse the risk so customers don't have to face the risk when they buy from them. On the other hand, traditional marketing is more often about taking. Traditional marketers expect customers to pay for everything – and they frequently expect customers to shoulder the risk."

Techno-savvy Solutions

While, traditional marketers tend to adopt technology slowly, Hepworth observes that entrepreneurial marketers quickly become very comfortable with technology as a valuable part of their marketing team, creating efficiency and capability at the same time.

The primary difference in approach – focussing on serving the customer as opposed to the traditional marketing focus on selling goods or services – shows up in many different ways.

"For example," Hepworth asserts, "traditional marketing is interruptive. It interrupts the customer with a daily barrage of messages, each focused on making the sale. Contrasting this approach, entrepreneurial marketers gain prospects' consent to send them useful information. They use opt-in mechanisms to broaden consent, before they try to get face-to-face with customers and prospects."

"Traditional marketing is also generally unintentional," adds the ambitious, seasoned CEO/Entrepreneur with a successful track record of starting, operating and selling services businesses. "It is mainly broadcast over mass media, reaching as many people who are totally disinterested and it reaches those who are interested. But entrepreneurial marketing is very *intentional,* being highly targeted to a small set of ideal buyers. Everything has a strategic objective, from the attire of the sales people to the way the phones are answered to the way sales people sell and all the content in every piece of public information. At the end of the year, traditional marketers count up dollars while entrepreneurial marketers count up relationships first, and dollars second. Entrepreneurial marketers know they can never make it too easy too much fun and too attractive to do business with them. But traditional marketers tend to do business the way most of their competitors do business, and as a result tend to be largely undifferentiated."

Direct Response and Quantum Thinking

Hepworth points out that while traditional marketers do a large amount of image advertising, entrepreneurial marketers never do image advertising "because they know it is almost impossible to measure. Instead, they'll do effective direct response marketing that will generate sales – with the image coming along for a free ride. It means thinking outside the box – but that's what entrepreneurial marketers do."

Another example of thinking outside the box, Hepworth adds, is the tendency for entrepreneurial marketers to engage in quantum thinking, a relatively new way of thinking, a holistic approach focusing on relationships and the interconnectedness of everything, with the whole being greater than the sum of its parts and with unpredictability a constant – the only true certainty being uncertainty – and the necessity to adapt quickly to a changing environment that's beyond anyone's control.

While entrepreneurial marketers nimbly response to changes in the marketplace, reinvent themselves and engage in new ways of doing things to achieve success amid change, traditional marketers are fixated on studying rulebooks that no longer apply. They're engaging in mechanistic thinking.

According to *Learning to Listen/ Learning to Teach* author Jane Vella, mechanistic thinking assumes a large degree of certainty, hierarchy and control in a universe that operates like a machine, It's a way of thinking that emphasises conformity over originality. Thinking outside the box, outside established parameters is discouraged; uniformity is desired; the whole is not greater than the sum of its parts; and to achieve future success, one should mirror established approaches that were successful in the past – without questioning whether they will work as well today.

Hepworth asks: "Are you an entrepreneurial marketer or a traditional marketer? Where are you on the continuum between entrepreneurial and traditional?"

"If you are positioned towards the traditional end of the continuum," he asserts, "recognize that you will need a big marketing budget to succeed. Entrepreneurial marketers will continue to chip away at your customer base."

"But if you are more on the entrepreneurial side, recognize that you need a constant flow of creative new ideas and lots of energy to make up for your lack of funds," he cautions.

"But remember, entrepreneurial marketers are above all patient. If you are feeling stress in your marketing, it is a sign that you are doing something wrong, so step back, take time to think, and recognize what you need to change."

Every business markets

One thing is certain: Every business markets – even if they think they do no marketing at all.

"The only really questions are what kind of marketing do you do – and is it effective, or not?" Hepworth notes.

"Yet I still come across people who proudly tell they do no marketing at all," he adds. "I hear this fairly often from professional services businesses. Many people find it easier to believe they don't market or to believe they have found the 'secret sauce' that means they don't need to market."

"But in fact, they do market," Hepworth explains. "I usually discover that like most people who believe they do no marketing, they rely predominantly on networking and referrals. To me, marketing is everything a business does to acquire and keep customers. So in my book networking and referrals definitely fall into marketing. I believe that these two tactics are often the default for people who have no marketing strategy or process. While these tactics can create quite a successful business, they eventually limit your growth because ultimately the amount of networking you can do is limited. When you are networking you can't deliver, and when you are delivering, you don't have time to network. If these are your only marketing tools, it will be extremely difficult to grow business beyond a few people."

Marketing can be low-cost – but not no-cost

Hepworth also finds that a lot of people who think they do no marketing base their mistaken belief on the erroneous assumption they are not incurring any marketing costs and any promotional efforts they make are free. "The truth is that networking and referrals have lots of hidden costs that are actually marketing costs. For example, think of lunches, dinners, coffee, membership fees, meeting attendance, your time, and travel and so on. Don't be fooled into thinking your time is free. It isn't. And remember there is also an opportunity cost to doing these things."

"By doing all these activities," Hepworth concludes, "costs mount up and you will be surprised to discover how much a new customer really costs you – even with seemingly free activities. Calculate how much you spend on the above items each year. Once you know all of your 'marketing' expenses for the year, divide that number by the number of new customers you get each year. I think you might be surprised at how high the number is. That's what you spend buying customers each year."

Hepworth says the danger in not understanding this is twofold: "First, there may actually be cheaper more efficient ways of marketing your business, but without any understanding of the costs, how can you tell? Second, if you think of your activities as free, you don't look at them as an investment in creating and keeping customers and this alters your behaviour on marketing and client service. I believe you take it less seriously than you should."

Attaining this crucial understanding often requires a change in mindset, he notes. "The first step is to recognize that you have invested considerable sums of money over the years of building your business. I believe marketing is simply a way of buying customers. For some people this may sound distasteful, but you should try to get used to it. If you get better at marketing than your competitors, then it becomes cheaper for you to acquire each customer than it does for your competitors and the more customers you get. What's more you can become more selective in the type of customers you acquire."

And with the change in mindset comes some interesting questions: "Once you see marketing and client acquisition as an investment," Hepworth observes, "it leads you to ask yourself: How do I protect that investment? How do I optimize that investment? Well, in order to protect that investment, you have to do everything to ensure that once acquired, a customer comes back regularly to buy more. This is also true in businesses where clients only buy once or occasionally."

Fall in Love with Your Customer

Hepworth counsels that one should not be shy about taking customer relations to a closer level. "Even in businesses where marketing is restricted to networking and referrals, client acquisition is still the most expensive thing you do. The best way to protect this investment is to fall in love with your customer rather than with your product or service. Falling in love with your customer will ensure you provide service that is without peer."

Too often, he notes, a business is love with its products and can't stop promoting them, when what is really needed is some customer attention, focusing on the customer's needs and wants and personal preferences. Taking this approach greatly improves the chances of gaining repeat business – and it sets you apart from many of your product-focused competitors.

"Being the same as everyone else in your industry is not protecting that investment," Hepworth asserts. "It makes you vulnerable to competition and over time erodes your investment.
Optimizing your investment in customers means finding ways to get each customer to buy as frequently as makes sense, and, every time they come back, offering them additional products and services that they need; the goal always being to serve the customer more fully. Almost everyone in marketing dreams of a

never-ending flow of qualified leads streaming into their business – and this is one way to get that stream flowing."

Hepworth has also done a number of informal surveys that show networking is a low-cost marketing tool of choice for many businesses for a number of reasons. "Based on what I heard, I have come to the conclusion that most like it because it doesn't really feel like work and in most cases it is non-pressured and non-threatening. It obviously works because some people have built successful businesses that way. But many are only mildly effective at it, resulting in a huge wasted opportunity."

He also points out: "StreetSmart Marketers constantly find ways to optimize every marketing dollar and every marketing activity. I too like networking, but am seldom content with meeting people and swapping business cards. This may seem like work, but it seldom yields anything other than modest results. Time is your scarcest resource and also the most perishable. You use just as much time working ineffectively as you do when working more effectively, but with some key networking skills, you can get a much greater return on your time investment. If you go to a networking meeting and only generate one lead per hour or generate five or six leads per hour your time investment is the same. I want to get the best return on my time investment I can."

Overcoming Common Mistakes

Reflecting on a recent business seminar he attended, Hepworth recalls the majority of people at his table gave ineffective introductory remarks. "It's a common mistake: Most people offered the most boring, traditional, undifferentiated introductions, some of them lasting several minutes. They started with their name and simply waffled on about all of the things they do. Most people lost me after their names and what they did. Their introductions were generally passionless, un-focused, made them sound generic, and in most cases, unless you're looking for those specific services, they'd have had no impact. Next time someone introduces himself in a business environment, listen and see whether you agree with me. Are you guilty of introducing your self in this way? What's it costing you if your introduction is like that?"

So why do so many business leaders waste their time and that of others on introductory comments that miss the mark? "They mistakenly think the broader the appeal the more likely they are to attract business," Hepworth suggests. "However, the opposite is true. People want to deal with specialists and experts. So the key to an effective introduction is to use something short, sharp and highly targeted that positions you as an expert. Thirty seconds is all you need, if you know who your target audience is and what you create for them. For example, a life insurance and investment advisor I know, introduces himself by saying he crash tests financial plans for thinking business owners. Most people are immediately interested and ask him how he does these crash tests. How many insurance salesmen get that kind of response? And how much more likely are you to get an interested response from an intriguing introduction like that one – rather than simply describing all the products and services you sell?"

"Another common mistake," Hepworth offers, "is not telling your contacts who you want to meet. The person you may be talking to is not always going to be a suitable prospect, but may be able to introduce some prospects to you. So my insurance friend could say something like; 'I want to meet business owners who are 10-15 years away from retirement, who want to know if their financial plans will give them the kind of retirement lifestyle they want. Do you know anyone who I might be able to help?' This approach gets the listener thinking of how they can help you build your business. They become a valuable resource on your behalf."

Rewards and Refunds

Hepworth has long held the practice of offering to send something of value to the people he meets. "This is often a piece of information that they might be interested in or find useful. Sending this gives me two contacts with the person in a short timeframe, creating the likelihood that I will be remembered, while providing a reason to stay in touch if I want to. It establishes me as giver, not simply a taker. A law of human nature is that people are more likely to reciprocate, if you first give them something useful."

Just as important as rewards or gifts, Hepworth notes, are promises of refunds if goods or services are not satisfactory.

"StreetSmart Marketers know networking is important and how to wring the maximum benefit out of every networking opportunity," he notes. "But here's something that's sometimes overlooked: Most of us would refund an unhappy customer's money without hesitation; why not make it an explicit benefit of your offer up front. Build it into your advertising, direct response and sales pitches. Make it obvious, instead of casually mentioning it or only offering when necessary. Reversing the risk and shouldering it yourself instead of expecting the customer to take the risk will build trust and encourage trial and increase sales."

Advantages outweigh Disadvantages

"Will customers take advantage of you?" he asks rhetorically. "Sure, some will, but the majority won't, and the risk reversal will increase significantly the number of buyers willing to try your service. Think of those that abuse you as a promotional cost. You'll find the advantages quickly outweigh the disadvantages."

Hepworth also advises making your guarantee powerful and appealing – the more outrageous the better. "I prefer to offer a better-than-money-back guarantee, where the customer gets to keep something of value, simply for trying the product or service, regardless if they keep the product or not. Customers should pay for your products and services, but you have got to make it easy for them to buy. This is one very effective way of doing it."

All of the advice Hepworth has shared in this chapter is only a small sample of the knowledge he shares in his *StreetSmart Marketer* book and electronic newsletter – both of which are considered required reading by an array of successful business executives, many of whom derive further benefit from engaging his consulting services. Indeed, sharing his expertise represents a deep well of satisfaction for this StreetSmart Marketer.

"This has been a lifelong quest for me. It has been a strange and exhilarating journey. Along the way, I've built four businesses and learned many valuable lessons; many of them the hard way, bumping my head as I went. If I can pass on some of what I've learned to others and help them succeed in business – that's something I feel really good about."

Leadership Lessons:

1. Ask questions to find out what your client or prospect is interested in or worried about, and then tailor your approach to address both.

2. Don't practice *Me Marketing* and try to impress a client or prospect by bragging. Practice *You Marketing* aimed at helping the customer. Make sure your marketing materials focus first on customer issues before your capabilities

3. Never pitch without first building rapport. Accept that many customers don't buy from you because you have the best products and services, but because you understand them as people. Get to know clients before selling to them.

4. Practice low-cost, low-risk, low-effort marketing methods that yield strong results. Realize that sometimes the most effective marketing approaches are also the least costly.

5. Learn from your failures as well as your success. Every time you come up short there is a new lesson to be learned.

6. Practice *Entrepreneurial Marketing* over Traditional Marketing for approaches better suited to the real world.

7. Make the easy sale first and then build on that success by straying in touch with your customer and meeting their needs on an ongoing basis.

8. Use an array of low-cost entrepreneurial marketing tools, including testimonials, joint ventures, strategic nurturing of prospects, referrals, customer education, public speaking, writing articles, and obtaining endorsements.

9. Fall in love with your customer, not your product/service.

10. When introducing yourself and your services, make your comments as intriguing and interesting as possible.

11. Give customers rewards, something of value, to help develop goodwill – and future business opportunities.

At a Glance:
Infinity Communications:

Leanne Bucaro

Title: Founder, President of Infinity Communications Inc, based in Oakville, Ontario, Canada and serving small to medium sized clients throughout North America and beyond.

Claim to fame: Infinity Communications was founded in 1995 with the goal of providing communication services to Business to Business clientele. Infinity Consultants have extensive experience working with the Fortune 100, Government and start-ups at an executive level. They are experts at building name recognition through focused programs that include media and analyst relations, executive visibility programs and consistent branding through clear and concise messages. Their capabilities extend to assisting the sales and marketing organization with effective messaging, strategies, process and skill development. There are two divisions in the company; one that caters to businesses that want Infinity to do the PR for them (Infinity Communications) and PR Mentor (www.pr-mentor.com) which is a new, affordable and simple way for small business to do it themselves, on-line. But, Infinity is available to guide their clients through the process. Either way they cater to the small and medium size market – clients choose to engage Infinity based on their specific requirements.

Financial Data: Privately held multi-million-dollar Company.

For More information:
Contact: Leanne Bucaro, Alan McLaren, Infinity Communications
Address: 69568-109 Thomas St., Oakville, Ont. Canada, L6J 7R4
Phone: 905-257-5555 or toll free: 1-866-841-2992 (US/ Canada)
FAX: 905-876-3866
Email: leanne@infinity.ca or alan@infinity.ca
Website: www.2infinity.ca and www.pr-mentor.com

7

Infinity Communications
"Live the Brand"

"At the end of the day your business is about selling and making customers happy – with the competitive realities of today you need to be firing on all cylinders. We help you do just that by setting a course for a consistent integrated approach to driving revenue."
- **Leanne Bucaro**, Founder, President, Infinity Communications

Live the brand! That's our motto – and in writing a book on how innovative leaders assist other businesses to become successful, I'd be remiss if I neglected to mention my own firm: Infinity Communications Inc.

At the risk of sounding immodest, I'm very proud of the success we've achieved to date in assisting other companies soar to greater heights in the global marketplace.

We've helped many companies effectively hone their message, communicate more clearly and grow their business through our integrated approach. Prior to writing this book, I was already quite familiar with the stories of the firms featured in these pages, because most of them are also Infinity clients.

First, a brief introduction is in order: Infinity Communications is a full service communications firm specializing in public relations, marketing and sales strategies for small to medium size companies in the B2B (Business to Business) space. Our mandate is to assist our clients with their business growth strategies through focused communication programs designed to build brand awareness and drive revenue growth.

Where I got started

I had started my career in the film and television industry on-air in front of the camera and eventually writing behind the scenes for TV. I subsequently transitioned into advertising and then moved once again into corporate communications with the Ontario provincial government. I made my way up the corporate ladder and excelled in various management positions.

One of my best decisions (and scariest) was to move into the high-tech market with Bell Mobility where my first project was to orchestrate the media launch of Digital PCS (I loved talking on my mobile – but had no idea how it worked). I also worked on the team that integrated and launched Bell, Bell Mobility and Bell Express Vu as an integrated retail store called Bell World.

I'm a member of many different Women in Technology Associations as well as an active member of the e-Woman's Network. I have been nominated for the Top 40 Under 40 Business award and served on the Prime Minister's Task Force On Women Entrepreneurs – and I absolutely love what I do.

How Infinity Communications Began

How I came to get involved in being a founder of Infinity is a story with more than a few twists and turns:
Infinity Communications was originally founded by me in 1995. I started my communications company as a way to help independent business build their reputation in the market.

Most of our initial business was logo design, communication planning and media relations strategies. The problem: I kept getting hired by my clients to work for them full-time. I loved my clients so much and became so involved; I couldn't imagine a work life without them. So Infinity Communications stopped and started, on and off, from 1995 to 2004.

During the tech boom of the late 1990s and into the current millennium, I was hired away (once again) – to a publicly-traded IT Security firm where I met Alan McLaren for the first time. This company quickly became media darlings in a world where people and a lot of businesses didn't know why they needed security.

People thought "hackers" were old newspaper guys. We had huge media success – but unfortunately – not as much revenue success as we did in building media value.

I had been working at this firm as Director of public relations and investor relations and, in early September 2001, was enjoying a rare vacation, a four-week stay in Europe that began in mid-August of that year.

While relaxing poolside at an Italian villa, I took a phone call and discovered the IT security company I worked for was going under and I wouldn't be paid salary and vacation pay owed to me. But they still hoped I'd attend a meeting at the NASDAQ stock exchange offices at the World Trade Centre in New York. The meeting was set for September 11.

Since the company wasn't going to pay me what they owed me, I told them there was no way I was going. And so, by the grace of God, I wasn't at the World Trade Centre on September 11, 2001 – but we lost a lot of friends that day; I was just walking around in shock, crying. It stunned me into silence.

But while the company I had worked for got caught in the high tech meltdown, 911 created a heightened societal security consciousness – including the need for enhanced computer and IT security throughout society, including computers and IT.

New Challenges

Alan and I found ourselves looking for the next challenge and little did we know what the future held for us at that time.

The next opportunity came quickly within three months we were asked to join a privately held IT security company as two of four partners – and together, Alan and I contributed to growing the company from about $1-million in revenue to $7.3-million in revenue in just three years. Our successful run was based on some excellent media success in building approximately $100-million worth of brand value for this firm.

For example, when the 'I Love You' virus hit computer systems, I rose at 5 am and quickly sent out advisory press releases, and then spent the rest of the day conducting interviews with CBC The National; CTV; CITY TV, numerous regional papers, radio stations, plus the Toronto Star, Toronto Sun, National Post and The Globe & Mail. I've also made frequent appearances on CP24, CTV NewsNet and CNN plus numerous industry trade publications. And I've arranged countless media interviews for our clients, raising their profiles and establishing them as industry experts.

The media exposure I'd obtained helped the sales team grow the business to the $7.3-million level – a seven-fold increase that was helped immeasurably by $100-million worth of publicity.

Alan McLaren

Alan had joined this IT security firm as company president in December 2001, a few months after I'd joined the firm. He was responsible for sales and vendor relationships as well as guiding and delivering on the vision of supplying mission critical security software and hardware solutions, with professional and educational security services to firms and governments worldwide.

He'd concentrated on extending and optimizing the companies' core competencies and operating excellence. He has an athletic background and drew on his natural abilities to lead by example while fostering a team spirit and environment. With decades of experience in corporate management, a strong background in sales, operations, marketing and information technology, Alan was a key member of the executive team.

But after three years with this company Alan and I were ready for new challenges, so we sold our shares and began to look for the next win. Our greatest success turned out to be an omen for our future; we just didn't know it then.

On leaving that IT security company, Alan was determined to find a role as president in a big firm. I had again resurrected Infinity and wanted to find opportunities with Infinity that we could pursue together. I said: "Sure Alan, go after a big corporate presidency – but in the meantime you can continue to help me get business for Infinity, and we'll work out a split of some sort. Alan said that until he found something, he would be happy to work together on this venture. Because we had worked together for almost five years, we had unwavering trust in each other – as well as the same morals and values.

Alan struggled with the idea of being a true entrepreneur, as he had a young family to support – starting over was initially not very appealing – yet he had a lot of great entrepreneurial skills – he just didn't know it at the time.

I really liked the partnership idea – but I had to allow Alan to come to the realization that we could do something special as business partners once again. But he was still hesitating. I knew ultimately that he would come around to my way of thinking. We were successfully winning clients – without hardly trying. But I refused to "push."

As we were winning clients, we began to shape our vision. We thought about what made our last company successful and how we saw SMBs (small and medium size businesses) being underserved by the larger PR and marketing firms.

Well, they say the way to come up with a better answer is to find a better question. The question was: "How come many SMB clients don't use PR as a basic strategy to differentiate their business?" The answer: Current suppliers could not speak their language as they hadn't lived the entrepreneur's life – of making payroll and marketing on a shoestring, with no financing etc. We had, and we knew how to do it. Now we needed to package it and use our marketing, PR and sales skills to take it to market.

Live the Brand Strategy

The secret? SMB clients cannot look at sales, marketing and PR in a vacuum. It is all about building a brand and a positive reputation and ultimately driving revenue. This "secret" drove us to our vision or "tag line" of "Live the Brand."

This simply means that every company needs to ensure that their messages are clear and concise and can be expressed by every person in the organization from the CEO to the Receptionist. As every successful company knows – Marketing is not a department, it is your business. It is about alignment and integration of Marketing, Sales and Public Relations.

With that foundation in place, the new fully mandated, fulltime Infinity Communications Inc. was born. And with that birth came a new addition to the Infinity Communications family – Alan was *in* – and I could not have been more pleased.

The catalyst was in shaping the strategy. The vision began to unfold and the opportunity to build another winner was very clear to both of us. With both of our families' support – we were off to incorporate the new business and get started in a big way.

As partners, we both brought different skills to the newly incorporated company. This allowed me to be creative and do the right brain thing, while Alan took a more methodical approach. Alan is very much a process guy and needs everything in boxes.

So what are we about? Choosing our name "Infinity Communications" had some thought behind it. Effective communications and branding has a beginning but does not have an end. It's an infinite process of refining and tweaking. Branding and effective communications is not a destination – it's an ongoing journey to brand relevance that's a focus of all companies' success strategies. Our name was chosen with that reality in mind.

Alan's Background

A large part of Infinity's new integrated approach is owed to my partner Alan, who has an extensive background in management, sales and marketing. In fact, Alan has more than twenty years of experience in corporate management, plus a strong background in sales and marketing, operations, HR and information technology. Alan is able to assist organizations in executing corporate strategy through the effective use of value activities for both management and sales. He's held the President title at many large companies – so he knows of what he speaks.

As well, Alan understands that each company has unique cultural challenges, but there are proven winning strategies and tactics that can be ingrained over time to improve the output of the sales organization. Alan has managed sales teams, developed sales and management training programs, run sales divisions and as President of IKON Quebec and a NASDAQ listed IT Security firm (where we met) Alan's executive business experience includes companies that range from private start ups to Fortune 50 public organizations. This perspective allows for a fresh, unique approach to each organization, by leveraging strengths and eliminating non value activities so that the focus on customers and profitability become key drivers for the sales organization.

Alan (who is bilingual) has been active with media engagements; speaking on topics that span from Identity Theft to Privacy and Security. He is often quoted in industry publications and has been interviewed on CP24 TV and CBC radio (International). Alan is a member of World President's Organization (WPO) Ontario. All of this, combined with decades of experience in corporate management, a strong background in sales, operations, marketing and information technology, makes Alan a great Infinity asset. Alan has been extensively media trained (the last five years by me) so his knowledge of the media game is extensive – and he looks at the PR world with different eyes. He often remarks "Just because it was done that way, doesn't mean we can't break the standard – let's look at things differently, so we can serve this SMB market effectively."

This experience also allows Alan to truly understand the challenges that leaders of business face every day, and it seemed logical that adding our two skill-sets together would make for an interesting and unique partnership.

Putting it all together

Starting your own business that helps people start their own business – or grow their existing one – is one huge exciting ball of stress punctuated by moments of stark raving terror.

I like to think I taught Alan how to listen to his gut – and he taught me one very important lesson: everything needs a process. Even creative right brain stuff needs a process. And that is where I think we have the competition beat: It is not just a case of "you should really do this – it would be too cool;" clients also need to see the process to understand it. The creative gal and process/box guy got to work putting a process around PR. We worked backwards, from the end, back. Our clients tell us we have the most comprehensive approach they have ever experienced.

Working with internal staff as well as external marketing support where appropriate, we help you distill your brand essence and communicate it consistently to all of your company's constituents. We believe it is as important to have your receptionist be able to communicate the company's message as it is for the CEO. Each customer "touch point" needs to be examined to ensure the company messaging is accurate.

Finally, we can work with your sales department to confirm that the messaging "on the street" is consistent. Many times we find the story is "adjusted" to the comfort level of the individual sales representative. But in order for your brand to be consistent throughout the company, this critical area must be addressed.

As for PR Services/Communication Planning: We begin all of our customer engagements with the creation of a comprehensive communication plan and checklist. Public Relations has many facets and nuances and it is our responsibility to maximize opportunities for our clients. The deliverable is a living breathing document that allows both Infinity Communications and our customers to be focused on the priorities established in the plan. Here are some of the key areas we work on with our clients.

Media and Key Message Training

Are you ready for your interview? Ready or not, you never know when the media will call. That's why it's best to be prepared. The key to successful interviews with journalists is to keep your message simple, interesting and newsy. We begin all of our client engagements with this first step as it is the foundation of an effective media strategy. You must have your messages honed and know how to deliver them – this is why we start with this training.

What you say and how you say it can have a lasting impact on your business because the media helps Canadians form opinions. Even if you are not actively seeking media attention, you never know when the media will call. So you need to be prepared.

What you will learn in our session: After a brief introduction to media relations, we use mock interviews to prepare you for print and/or broadcast (radio and TV) interviews. The interviews take place in real time and are taped. You receive immediate feedback after each interview. By the end of the seminar, you will be able to present your message (and yourself) in a professional and positive manner. You will understand: Your rights before, during and after the interview; setting key messages and a message agenda; presenting your key message points from the first question on; how to keep your message positive even when responding to negative questions; the implication of "off-the-record" comments; and, how to maintain your composure, even if ambushed.

Media Strategy

At the end of the day it's all about seeing your name in print or in other forms of Media (TV, radio, speaking engagements, etc). At Infinity Communications, we specialize in public awareness and corporate exposure through media and analyst relations. Everything we do revolves around communicating your name and your key messages through the proper media channels to reach the people who make the decisions whether to buy your product or service or go with your competitor. We understand media, having spent years working in print and broadcast media. We know what reporters, editors and producers need. And we've built solid working relationships with local, national and international media.

Executive Visibility Program

Infinity's Executive Visibility Program (personal branding) is a huge plus for many clients: Business leaders recognize the need for effective public relations to help them stand out in the crowd. Through carefully crafted executive visibility programs, Infinity will help you establish yourself as an industry visionary and technology leader. Whether it's booking and writing the keynote address for the right trade conference, or arranging a CEO dinner with analysts and journalists, Infinity builds positive reputations for companies and their leaders.

Infinity Communications Branding

We believe that Branding and Communication extend well beyond the Marketing Department. Your Brand *is* your company. And Marketing is the communication of your brand to the outside world. This applies whether you have the resources of a fortune 500 organization or you are a small or medium sized business.

The challenge is communicating those messages effectively and consistently to your target audiences. Branding is more than your logo, it is more than a tag line and it is more than a well crafted advertising campaign. It is the personality and core value proposition you present to your customers. Your value proposition and persona need to be reflected in all elements of your marketing mix. Brand is what people think about when they hear your name – it is the perception of your company in the market.

In our view "the Brand" or your brand is affected positively or negatively at every "touch point" you have with your prospects, suppliers, customers, potential recruits – every point you reach outside of your organization. In other words, everything you communicate with the marketplace impacts on your brand.

We believe that to "Live the Brand" you need an Integrated Thinking approach to your brand strategy. It does not start and stop with Marketing. It involves every employee, supplier, partner, and so on, with whom your business comes into contact. It can be as simple as ensuring your receptionist can communicate your core messages and/or how you deal with individuals looking for employment at your company. Everyone that "touches" your organization in any way must come away with a good positive feeling. This enhances the essence of your brand. And this is why "Live the Brand" was chosen as our mission for Infinity Communications. We want to help our clients "Live their Brand."

Can You Afford Public Relations?

Can you afford Public Relations for your company? It's a question we sometimes hear from struggling companies. We advise them to get the answer by asking themselves these questions: Did you ever lose an account where you had the better value proposition? Maybe you were beaten by a better brand? Do your sales executives tell you "no one knows who we are?" Do they spend a lot of time on ineffective prospecting? Do you rarely get leads because the prospect has never heard of you?

The cost of PR and branding can be summarized this way: If you needed marketing people at x dollars per year – you would hire them. PR and related services can be equated to the cost of an employee per year without the sick days and management issues related to a staff member. Additionally, when you compare the effectiveness of PR versus advertising, you note that advertising is in your voice, as you are selling something; whereas, a PR message in the media comes with an implied third party endorsement, the media. Which do you think has more sticking power?

As well, the brand can sell for you when you have an integrated approach to building your brand. A good brand shortens the sales cycle because your credibility is established early, which saves time in the prospecting phase. Your sales people can actually increase their productivity with an investment in your brand.

Intense Focus on the Customer

At the core of every successful organization is an intense focus on the customer. The challenge is to service the customer while providing the profitable revenue that all stakeholders demand. Often there is misalignment between the core mission of the organization and the "feet on the street." Successful organizations close that gap.

Many organizations poised for the next tier of revenue and profitability growth need assistance getting there. As organizations mature, the "we've-always-done-it-that-way" method is not as effective as it once was. Competition, commoditization of offerings and other factors impact the bottom line.

We recommend strategies and tactics based on proven best practices that will impact on growth. When you have explicit expertise outsourced you can focus on your core business – it is critical for the leaders to continue to work "on the business" not "in the business" so that future growth can be achieved.

Essentially, we are the eyes and ears from the outside – our perspective is not clouded by internal politics and challenges. We help you to do the things you know you need to do to achieve growth. We make sure that all agreed-upon tactics are completed with efficiency and effectiveness.

So what does Infinity Communications Inc. look like today? Our Infinity "Live The Brand" Strategy takes an integrated approach combining Public Relations – including communication planning, key messages, media training, media relations, executive visibility and special events; Marketing – including strategic planning, consulting, implementation, website creation and design services; and, Sales – including business development strategies, sales structure, sales process, sales effectiveness and coaching.

Our Partnership Works for You

So what about the Infinity partnership? Our advice is simple – If you cannot be 100% sure that your values are aligned – don't have a partnership. In many ways it is like a marriage, so trust is paramount. We have been very fortunate to have built a foundation of trust and integrity and we live our business lives on this foundation. Learning to work together in our new Infinity Communications company was easy because we knew where the strengths of one partner start and the other begins, so we did not try to be what we could not be. We let the talent flow and learned from each other – as we continue to do today.

It is also not about ego. We don't care who gets the credit – we just care that our clients are getting value.

Finally, we laugh everyday – whether in person, email or phone – we have a blast. We laugh at each other with each other and with our customers. If you cannot do that – why bother.

We are two years into this wonderful experience called Infinity Communications Inc. We are expanding our business, having fun with our clients and each other.

We're also building a company that will change the way Communication firms deal with SMB clients.

And we invite you to give Infinity the opportunity to help your company achieve new levels of success.

Oh, and don't forget to "Live your Brand!"

Leadership Lessons:

1. Learn from your failures and mistakes as well as from your successes to construct a winning approach to business. You'll make a lot of them if you're *really* trying to grow.

2. Gain wisdom from business battles. Employ this expertise to become more effective at whatever you do.

3. Don't try to reinvent the wheel. Sometimes it's far better to hire an expert rather than attempt to become one by learning an unfamiliar process. Your time is too valuable to get sidetracked away from building your business.

4. Set your strategy and make it simple: One page stating how you will get to Stage Two – and the steps needed.

5. Hire someone either internally or externally to be your communications arm.

6. Make sure the world knows how and why you are different from your competitors. Media exposure is worth ten times what an ad is worth. Who reads ads anyway?

7. Set realistic expectations of yourself and others and live up to them. It's counterproductive to expect more than that.

8. Establish your firm's structure before its founding date. Clearly set out duties, responsibilities for each position.

9. A quote we live by: "You only see obstacles when you take your eyes off your goals." Set goals, write them down.

10. Consider the benefits of strategic alliances for providing you with expertise and resources you might not have.

11. Take an hour a week to work *on* your business. The only way to grow your business is to think about it regularly.

12. Keep it simple. Do five things 4,000 times rather than trying to do 4000 things. And make a *stop-doing* list. If it doesn't match with the plan – don't do it.

Manor House Publishing Inc.
www.manor-house.biz 905-648-2193

Leaders
The Complete Summary
Of Leadership Lessons

WSI Leadership Lessons:

1. Find an untapped market and service it. This is the approach WSI has taken to providing Internet expertise to small to mid-sized enterprises. Prior to WSI, this degree of expertise and service was normally only available to large IBM type client firms. This lesson, and the others listed here, can be applied to numerous other marketplaces. Reaching out to overlooked markets can yield rich dividends for you and your clients.

2. Create and develop ways to provide the highest quality products and services to your customers at the lowest possible costs to overcome client affordability issues and capture market segments previously lost due to client cost concerns. This approach inevitably generates increased business volumes, which in turn allow for increased production, bulk buying and economies of scale that enable reasonable pricing and significant profits.

3. Develop a highly effective website to fully promote and sell your products and services to a vast marketplace. Be sure the search engines can bring visitors to your site and ensure your site is attractively presented and user friendly.

4. Make your Internet presence as strong as possible, fully utilizing the Internet and your website as a tireless sales force operating 24 hours a day, seven days a week to raise your profile and generate sales revenue on your behalf.

5. Give your website full e-commerce and e-marketing capability to drum up as much business as possible. Realize that the costs of hiring WSI consultants and putting needed systems in place is minor compared to the benefits. Realize too that the full costs involved are often less than an advertisement in the phone book and that every day that you don't have such capabilities in place is costing you money in terms of lost sales and business you've failed to capture.

6. Achieve success by helping others achieve success. Position your company as the go-to firm for expert advice and superb products and services that help other firms succeed.

7. Don't just sell products and services – assess each client's needs and match them up with those products and services to best suit their needs. Adopt a long-term approach that ensures whatever you sell a client today can be added on to tomorrow as the customer's business grows and requires updated and expanded products and services.

8. As you achieve success in a financial sense, strive to achieve equally satisfying success in helping the less fortunate help themselves. WSI's work with World Vision fighting child poverty and revitalizing Third World villages provide good examples of this approach.

9. Be there when your customers need you. Don't simply make a sale and walk away, be available to provide follow-up assistance and advice. This approach is also crucial for gaining trust, developing repeat business, follow-up sales and referrals.

10. Accept and embrace opportunity. When opportunity knocks, open the door. Develop goals based on something you're good at and that you enjoy doing. Play to your strengths. Think about what you want to achieve and set about working hard and smart to make your dreams a reality.

Robert Kubbernus Leadership Lessons:

1. Realize that there are no rules, just guidelines, and every situation is different; calling for tailor-made solutions.
2. Get rid of fear: If you think you can't succeed it's a self-fulfilling prophecy. Rise above fear and seize opportunity.
3. Challenge yourself to do more: The more tasks and challenges you take on, the better you become and the more you raise your threshold for pain and exhaustion. Simply put: The more times you take on the tough stuff, the better you get at it and the more you enjoy doing it.
4. Deal with the "tough stuff" – the difficult decisions – first and save the easy problems for last when you've got your business back on the right track.
5. Know when to reach out for help and don't be afraid to do so. Your own abilities and persistence can often take you most of the way but few people succeed entirely on their own. Help is often available when you most need it and it's amazing what you can get simply by asking for it.
6. Don't think that if you ignore a situation it'll eventually heal itself. It's more likely to get worse over time and become that much more difficult to solve as a result of your earlier inaction.
7. Don't hesitate on initiating a necessary decision or course of action simply because it's uncomfortable or distasteful. Delay simply adds to the difficulty in doing what must be done. Better to act fast, take the distasteful medicine, and get it over with. As Kubbernus asserts: "If you have to swallow a frog, it's best not to stare at it for too long."
8. Force yourself to solve problems and compare your situation with that of others and study how they resolved their difficulties. You'll be a better business person.
9. Always look for the unexpected whenever problems arise that seem to defy solution. Go over the details and find the missing pieces. The answer is in the problem itself.
10. Know your limitations in terms of ability. There's no shame in hiring experts to help – in fact it's a good idea.
11. Don't over-extend yourself – make sure your business is well financed and growing at a sustainable rate.

Sam Mercanti Leadership Lessons:

1. A good leader leads by example, builds and empowers other leaders; provides direction without micromanaging.

2. Praise employees for what they do right; accept that they make mistakes and help them correct and learn from them.

3. Not everyone wants to lead: Do not force leadership on anyone, but give opportunity to those who do wish to lead. Groom the next generation of leaders for challenges ahead.

4. Get in touch with customers: Get out on the shop floor or the street to get a feel for how your company is perceived.

5. Surround yourself with talented people and draw on their expertise to win business battles.

6. Adopt strong values and ethics and live up to them.

7. Establish and nurture relationships with other people. Treat others as you would like to be treated.

8. Don't die with your dreams still in your heart – make your dreams a reality.

9. Try to be all that you can be. It means hard work but the results are well worth it.

10. Don't deny yourself the pleasures of time with loved ones.

11. Take calculated risks; there is no reward without risk.

12. Go out and achieve – give life your best shot.

13. Internalize a belief system that has confidence in your ability to conceive, believe and achieve.

14. Don't just satisfy your customers. Delight them by going the extra mile and delivering added value.

15. Believe in yourself. We are capable of doing whatever we put our minds to. Grow each day in mind, body and spirit.

Manor House Publishing Leadership Lessons:

1. Don't make assumptions: Take a proactive approach with clients to determine their needs and define your services.

2. Rather than start from scratch, forge alliances to tap into the expertise, know-how and resources of complementary firms and benefit from their years of experience.

3. Don't be afraid to make changes. There's always room for improvement. Small changes can make a big difference.

4. Let publishers deal with printers, bookstores, distribution. Self publishing is often a recipe for headaches with hidden or unforeseen costs. Unless all efforts fail, use a publisher.

5. Consider writing or being the subject of a book to establish yourself as an expert. Differentiate yourself from rivals.

6. Think outside the box and develop concepts that work better for you and your customer. Use your imagination.

7. Instead of forming a partnership, which can restrict your independence, hire freelancers and form alliances to get services you need while you alone control your business.

8. Once you have mastered one area of your business, expand into others to provide you with multiple income streams.

9. Train yourself to see potential where others do not. Make any modifications needed to help a flawed project succeed.

10. Utilize consultants to gain expert understanding/advice on key issues, rather than spend an excessive amount of your time trying to attain the same answers on your own.

Mariposa Cruises Leadership Lessons:

1. Always look at opportunities, assess them, and determine if they're doable – and then see if you can reduce the risk in some way while still deriving benefit.

2. Taking risks is often necessary to achieve success, but make sure that each risk you take on is a calculated risk in which the odds of success are in your favour.

3. Leadership involves being the best in several key areas. For example, Mariposa Cruises offers the highest possible level of service on the best possible cruise ships in the best possible location.

4. Apply targeted marketing to wisely aim your advertising dollars directly at those most likely to use your services.

5. Leadership is also about translating ideas into action, communicating context, taking leadership, and creating innovative solutions as a team.

6. Providing great customer service is the surest way to build your business through referrals and become a leader in your field.

7. If undertaking a leveraged buyout of another company, strive for vendor take-backs of mortgages or debt to minimize costs and risks.

8. Increase your marketing to grow your business.

9. Partner with other companies that complement your own and derive mutual benefit from the synergies that result.

10. Don't cling to an outdated or ineffective message; re-brand your company to convey a clear message in a memorable manner.

The StreetSmart Marketer Leadership Lessons:

1. Ask questions to find out what your client or prospect is interested in or worried about, and then tailor your approach to address both.

2. Don't practice *Me Marketing* and try to impress a client or prospect by bragging. Practice *You Marketing* aimed at helping the customer. Make sure your marketing materials focus first on customer issues before your capabilities

3. Never pitch without first building rapport. Accept that many customers don't buy from you because you have the best products and services, but because you understand them as people. Get to know clients before selling to them.

4. Practice low-cost, low-risk, low-effort marketing methods that yield strong results. Realize that sometimes the most effective marketing approaches are also the least costly.

5. Learn from your failures as well as your success. Every time you come up short there is a new lesson to be learned.
6. Practice *Entrepreneurial Marketing* over Traditional Marketing for approaches better suited to the real world.

7. Make the easy sale first and then build on that success by straying in touch with your customer and meeting their needs on an ongoing basis.

8. Use an array of low-cost entrepreneurial marketing tools, including testimonials, joint ventures, strategic nurturing of prospects, referrals, customer education, public speaking, writing articles, and obtaining endorsements.
9. Fall in love with your customer, not your product/service.

10. When introducing yourself and your services, make your comments as intriguing and interesting as possible.
11. Give customers rewards, something of value, to help develop goodwill and future business opportunities.

Infinity Communications Leadership Lessons:

1. Learn from your failures and mistakes as well as from your successes to construct a winning approach to business. You'll make a lot of them if you're *really* trying to grow.

2. Gain wisdom from business battles. Employ this expertise to become more effective at whatever you do.

3. Don't try to reinvent the wheel. Sometimes it's far better to hire an expert rather than attempt to become one by learning an unfamiliar process. Your time is too valuable to get sidetracked away from building your business.

4. Set your strategy and make it simple: One page stating how you will get to Stage Two – and the steps needed.

5. Hire someone either internally or externally to be your communications arm.

6. Make sure the world knows how and why you are different from your competitors. Media exposure is worth ten times what an ad is worth. Who reads ads anyway?

7. Set realistic expectations of yourself and others and live up to them. It's counterproductive to expect more than that.

8. Establish your firm's structure before its founding date. Clearly set out duties, responsibilities for each position.

9. A quote we live by: "You only see obstacles when you take your eyes off your goals." Set goals, write them down.

10. Consider the benefits of strategic alliances for providing you with expertise and resources you might not have.

11. Take an hour a week to work *on* your business. The only way to grow your business is to think about it regularly.

12. Keep it simple. Do five things 4,000 times rather than trying to do 4000 things. And make a *stop-doing* list. If it doesn't match with the plan – don't do it.

Manor House Publishing Inc.
www.manor-house.biz 905-648-2193

www.ingramcontent.com/pod-product-compliance
Lightning Source LLC
Chambersburg PA
CBHW031253290426
44109CB00012B/564